Renewable Energy System Design

Renewable Energy System Design

Ziyad Salameh

Dept of Electrical and Computer Engineering
University of Massachusetts-Lowell
1 University Ave
LOWELL 1854

AMSTERDAM • BOSTON • HEIDELBERG • LONDON
NEW YORK • OXFORD • PARIS • SAN DIEGO
SAN FRANCISCO • SINGAPORE • SYDNEY • TOKYO
Academic Press is an imprint of Elsevier

Academic Press is an imprint of Elsevier
225 Wyman Street, Waltham, MA 02451, USA
525 B Street, Suite 1800, San Diego, CA 92101-4495, USA
32 Jamestown Road, London NW1 7BY, UK
The Boulevard, Langford Lane, Kidlington, Oxford OX5 1GB, UK

Library of Congress Cataloging-in-Publication Data
Application submitted

British Library Cataloguing in Publication Data
A catalogue record for this book is available from the British Library

ISBN: 978-0-12-374991-8

For information on all **Academic Press** publications
visit our web site at store.elsevier.com

Working together
to grow libraries in
developing countries

www.elsevier.com • www.bookaid.org

Contents

CHAPTER 5   Emerging Renewable Energy Sources 299

Preface

Constantly growing demand for energy that relies only on fossil fuels cannot continue indefinitely. The Earth's finite supply will eventually be exhausted. Energy is a major key to industrial development and the world's well-being.The awareness of the depletion of fossil fuel resources has challenged scientists and engineers to search for alternative energy sources that can meet demand for the near and distant future. It has been recognized that attention should be paid to resources that are continuous, renewable in their availability, and pollution free. This book is dedicated to an understanding of these sources.

Renewable Energy Systems and Design is intended as a textbook for a course on alternative energy sources for seniors and first-year graduate students in electrical and mechanical engineering. In addition, it could be used as a reference for practicing engineers involved in the design and application of renewable energy sources. The prerequisite courses are basic electrical circuits, physics, chemistry, and mathematics, typical of senior engineering students. The content of the book is designed for a one-semester course.

The book's purpose is to provide important knowledge to the reader about all aspects of renewable energy sources for a proper understanding of the present and future of them. The basic subjects are introduced in a practical manner and concepts are covered in considerable depth and detail. Once the reader has learned the information in the book, he or she will consider applying these sources in their practical life.

This book is organized into five chapters. Chapter 1 includes an introduction, which is dedicated to the causes of the enormous interest in renewable energy sources. Chapters 2 and 3 are dedicated to the most widely used renewable energy sources: Chapter 2, photovoltaic, and Chapter 3, wind energy conversion systems,. These two subjects are explained in more detail and emphasis is given to them, including design and analysis.

Chapter 4 is dedicated to energy storage, which plays a significant role in the application of renewable energy sources. It includes discussion of batteries, fuel cells, flywheel, superconductors, supercapacitors, hydropump storage, and compressed air storage. Chapter 5 is dedicated to emerging renewable energy sources; covered are OTEC, tidal power, wave power, biomass, geothermal, thermal energy conversion, and satellite power.

Factors Promoting Renewable Energy Applications

1.1 INTRODUCTION

Recently global warming, pollution, and high oil prices forced politicians, utility companies, and the general public to pay more attention to renewable energy sources (RES) such as wind power, photovoltaic, and biofuels.

RES are environmentally friendly sources of energy; they form the backbone of distributed generation systems. Interest in them during the recent decade has been reflected in many academic studies, which provide more information on new ways of using renewable energy in addition to the traditional, isolated, stand-alone systems.

The term distributed generation systems (DGS) refers to a system that consists of many smaller generating plants located close to the major loads, as opposed to a few large centrally located power stations. A distributed generation system can significantly reduce transmission loss since the power source is located physically closer to the load. Other advantages of DGS are the increased reliability of the power supply because there is less dependence on a few major transmission lines over a large geographical area and reduced capacity of the main transmission lines and transformers.

In today's world, there are many reasons that support enormous interest in renewable energy and the move away from conventional energy sources; the following are the main reasons:

- Higher oil prices—oil prices are going up and up because demand exceeds supply.
- The impact of fossil fuel usage on the environment is manifest in:
  - Air pollution from transit systems, transport of necessities, automobiles, and power stations has made cities hazardous to our health, especially to children
  - Acid rain created when nitric oxides and sulfur oxides combine with water in clouds
  - Global warming caused by the greenhouse effect
- National security can be affected if supply is iterupted due to a political crisis

Because of the competitive nature of the global market, availability of energy supplies is unpredictable. Figure 1.1 shows the pollution as a result of burning fossil fuel.

Atmospheric concentrations of carbon dioxide and CH_4 have increased by 31% and 149%, respectively, above preindustrial levels since 1750. Globally, the majority of anthropogenic greenhouse gas emissions arise from fuel combustion. Around 17% of

FIGURE 1.1 Pollution from the fossil fuel economy.

emissions are accounted for by the combustion of fuel for the generation of electricity using the conventional electric power plants, especially the thermal power stations.

1.2 US AND WORLD ENERGY USAGE FOR ELECTRICITY GENERATION

Figure 1.2 shows the percentage of electricity generated by fuel type in the United States. Out of the 3883 billion Kwhs generated, coal was the primary fuel type (~50%) used. The figure also shows that hydroelectric systems produce about 6.5% and nuclear stations produce about 19.3%, while use of renewale energy produces about 2.3%. The total energy available is close to 4055 billion Kwh.

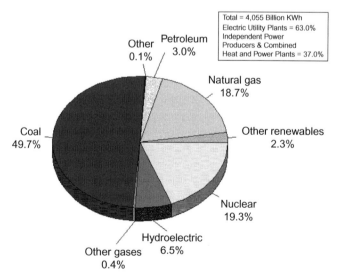

FIGURE 1.2 US electricity generation by fuel type as of 2005.

By comparing the electricity generation data (see Figure 1.2) to the projected electricity generation in 2030 (Figure 1.3), you get a good idea of the trends we will see over the next 20 years. Overall electricity generation will increase due to our expanding economy and population growth. This expansion will increase dependence on fossil fuels since the contributions from nuclear and renewable power sources remains relatively flat. Hydroelectric power also contributes to this because it remains stagnant, as can be seen in Figure 1.4. It is also clear that the share of renewable energy is on the rise and that the percentage of petroleum is not expected to increase due to the expected rise of oil prices. The share of coil will increase.

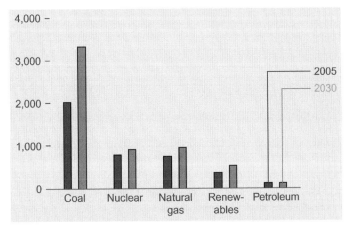

FIGURE 1.3 Electricity generation by fuel for 2005 and 2030 (billion Kwh).

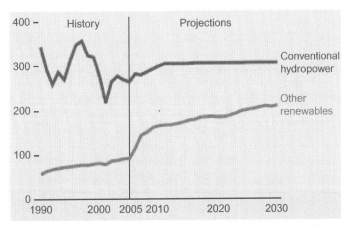

FIGURE 1.4 Renewable energy electricity generation, 2005 to 2030 (in billions of Kwhs).

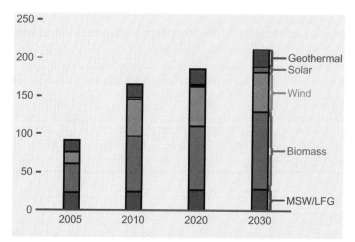

FIGURE 1.5 Nonhydroelectric renewable electricity generation (in billions of Kwhs).

If you then take a closer look at the electricity generated by nonhydroelectric renewable sources (Figure 1.5), you can see the future outlook of the technologies that currently exist. Solar power is not heavily used today, but it should see some growth over the next 20 years. Wind and biomass should show the most growth, while geothermal, municipal solid waste (MSW), and acronimous (LSW) will grow over the next few years and then level out.

The previous figures focused primarily on the US generation of electricity. Figure 1.6 shows the world's electricity generation by fuel type; it tracks the US

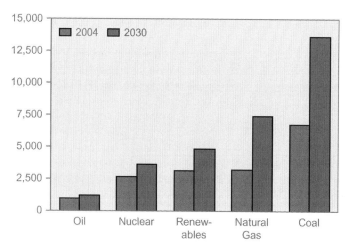

FIGURE 1.6 World electricity generation by fuel type, 2004 to 2030 (in billions of Kwhs).

figures fairly well. The use of fossil fuels (i.e., oil, natural gas, coal) continues to grow faster than renewable and nuclear power sources.

1.3 FACTORS AFFECTING THE APPLICATION AND USE OF ALTERNATIVE ENERGY SOURCES

1.3.1 The impact of high oil prices

Fifty years ago, the United States was self-sufficient in its supply of petroleum. Today, it imports more than half of its petroleum and consumes 25% of the world's supply. Until the mid-1950s, coal was the world's foremost fuel, but oil quickly took over. Following the 1973 and 1979 energy crises, there was significant media coverage of oil supply levels. This brought to light the concern that oil is a limited resource that will eventually run out, at least as an economically viable energy source.

Some would argue that because the total amount of petroleum is finite, the dire predictions of the 1970s have merely been postponed. Others argue that technology will continue to allow for the production of cheap hydrocarbons and that the Earth has vast sources of unconventional petroleum reserves in the form of tar sands, bitumen fields, and oil shale that will allow for petroleum use to continue in the future. For example, Canadian tar sands and US shale oil deposits represent potential reserves that match existing liquid petroleum deposits worldwide.

Today, about 90% of vehicular fuel needs are met by oil. Petroleum also makes up 40% of the total energy consumption in the United States but is responsible for only 2% of electricity generation. Petroleum's worth as a portable, dense energy source that powers the vast majority of vehicles, and as the basis of many industrial chemicals, makes it one of the world's most important commodities. About 80% of the world's readily accessible reserves are located in the Middle East, with 62.5% coming from the Arab countries: Saudi Arabia (12.5%), United Arab Emirates, Iraq, Qatar, and Kuwait. The United States has less than 3%.

Oil production

The United States produced enough oil to supply its own needs until 1970 at which point it had to start importing oil to meet the demand. Production in 2000 averaged 5.8 million barrels per day of crude oil; for 1999 was 5.9 million barrels per day. After the oil price collapse of 1985–1986, US oil production declined dramatically. Oil production in 2000 was down by 24% from 1985. However, according to the Energy Information Administration (EIA), in 2001 production increased by 70,000 barrels per day, or 1.1%.

There is little to no chance of discovering any significant new onshore oil fields in the United States. The possibility of discovering major deposits of oil offshore is good, but offshore drilling has been banned in many areas. There are several good prospects far offshore that are open to exploration; however, they are usually in very deep water and therefore extremely expensive to drill. The United States produces 12% of the world's oil and is concentrated onshore and offshore along the Texas–Louisiana Gulf

Coast. The area extends inland through west Texas, Oklahoma, and eastern Kansas. There are also significant oil fields in Alaska along the central North Slope.

The US demand for oil is increasing slightly every year, but domestic oil production is decreasing. The United States consumed an average of **19.6** million barrels per day of oil in 2000. It is estimated that it imported 10.9 million barrels of oil per day during the first eight months of 2000. At this rate, the United States is imports about 57% of the oil that is being consumed.

The main suppliers of oil to the United States at this time are Canada (1.68 million barrels per day), Saudi Arabia (1.49 million barrels per day), Venezuela (1.46 million barrels per day), and Mexico (1.35 million barrels per day). The United States has energy sanctions against Iran, Iraq, and Libya, all major oil producers, which prohibit US companies from doing business with them.

In August 2005, the International Energy Agency reported annual global demand at **84.9** million barrels per day (mbd), which translates into more than 31 billion barrels annually. This means consumption is now within 2 mbd of production. At any one time, there are about 54 days of stock in the Organisation for Economic Cooperation and Development (OECD) system plus 37 days in emergency stockpiles. The Association for the Study of Peak Oil and Gas (ASPO) was founded in 2000 by the geologist Colin Campbell. ASPO has calculated that the annual production peak of conventional crude oil occurred in early 2004.

During 2004, approximately 24 billion barrels of conventional oil was produced out of the total of 30 billion barrels; the remaining 6 billion barrels came from heavy oil and tar sands, deep water oil fields, and natural gas liquids (Figure 1.7). In 2005,

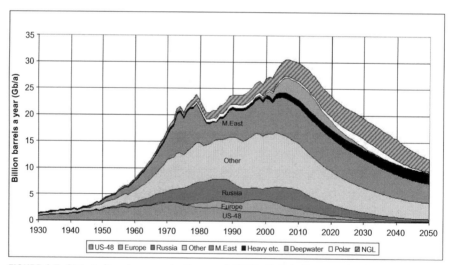

FIGURE 1.7 Global oil and gas production as of 2004. Global oil production is rapidly approaching its peak, even if natural gas liquids and expensive, destructive, risky deepwater and polar oil are included.

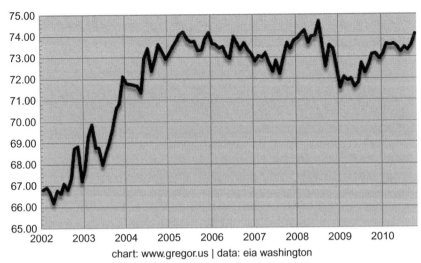

chart: www.gregor.us | data: eia washington

FIGURE 1.8 World crude oil production in millions of barrels per day, 2002 to 2010.

the ASPO revised its prediction for the peak in world oil production, from both conventional and nonconventional sources, to the year 2010. Natural gas is expected to peak anywhere from 2010 to 2020 (Bentley, 2002).

World oil production essentially has been flat since the beginning of 2005, as shown in Figure 1.8. Colin Campbell of the Association for the Study of Peak Oil and Gas has calculated that the global production of conventional oil peaked in the spring of 2004 albeit at a rate of 23 GB/year. Of the three largest oil fields in the world, two have peaked. Mexico announced that its giant Cantarell Field entered depletion in March 2006, as did the huge Burgan Field in Kuwait in November 2005. In April 2006, a Saudi Aramco spokesman admitted that its Ghawar Field, the largest oil field in the world, is now declining at a rate of 8% per year, and its composite decline rate of producing fields is about 2%. New drilling in Saudi Arabia may be able to replace some of the loss from the Ghawar Field.

Nor is the crisis restricted to petroleum. Traditional natural gas supplies are also under the constraints of production peaks, which especially affect specific geographic regions because of the difficulty of transporting the resource over long distances. Natural gas production may have peaked on the North American continent in 2003, with the possible exception of Alaskan gas supplies, which cannot be developed until a pipeline is constructed. Natural gas production in the North Sea has also peaked. UK production was at its highest point in 2000, and declining production and increased prices are now a sensitive political issue there. Even if new extraction techniques (e.g., coalbed methane) yield additional sources of natural gas, the energy returned for energy invested will be much lower than traditional gas sources, which inevitably leads to higher costs to natural gas consumers.

Oil consumption

The worldwide consumption of oil continues to increase each year. One of the main factors in the increase is the rise of China and India as global powers. As these countries continue to grow and invest in technology, their consumption of oil increases. The US Energy Information Administration reports that Chinese oil consumption has increased more than 40%. That increase has accounted for more than one-third of the total growth in worldwide oil demand over the past few years.

China became a net oil importer in 1993. China now consumes **7** million barrels per day, half of which is imported; it is the world's second largest oil consumer and the third largest oil importer, behind the United States and Japan. Nearly half of China's imported oil comes from the volatile Middle East, where it has been busily signing long-term supply deals, including a $70 billion oil and gas deal with Iran. China's future trends are even more dramatic than those of the past.

Following the recent surge in China's demand for petroleum, its per capita consumption of oil is still less than 8% that of the United States. Adding 5 million cars per year, China expects its auto fleet to increase from less than 25 million today to more than 125 million within 25 years. As a result, the EIA projects that China's demand for oil will more than double by 2025, reaching 14.2 million barrels per day, with more than 10.9 million barrels being imported. As a result, China's net imports will have increased by 8 million barrels per day since 2004.

Until former President Bush joined his six immediate predecessors by promising that the US addiction to oil would end, the EIA had been projecting that net petroleum imports, which averaged 4.2 million barrels in 1985—when China was East Asia's largest oil exporter—would exceed 19 million barrels a day in 2025, reflecting an increase of more than 7 million barrels since 2004. The trends and mathematics point to a significant increase in the possibility for problems to emerge, almost certainly in the Middle East.

Figure 1.9 shows the trends in world energy consumption from 1980 and makes a prediction for 2030. As can be seen, the world consumption of all forms of energy is expected to increase and has been rising steadily since the 1970s. Oil consumption is the largest form of energy used worldwide.

Lately the oil demand of the world exceeds oil production, as shown in Figure 1.10. Figures 1.11 through 1.14 display relative oil production and oil demand for various countries throughout the world. It is well known that the United States is the largest consumer of oil in the world, followed by China, United Kingdom, and India. Oil consumption is in red and Production is in black.

The cost of oil

The price of standard crude oil on NYMEX designated contract market was under $25 per barrel in September 2003. By August 11, 2005, the price was above $60 per barrel for more than a week and a half. A record price of $75.35 per barrel was reached, due in part to Iran's nuclear crisis, on April 21, 2006. Although oil prices were considerably higher than before, they were still roughly $14 from exceeding the inflation-adjusted "peak of the 1980 shock, when prices were [more than] $90 [per] barrel in today's prices" [35]. In 2008, the price of oil spiked to $145 per barrel.

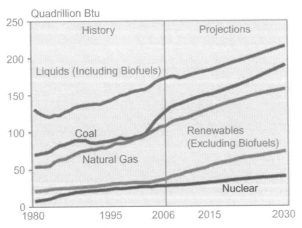

FIGURE 1.9 World energy consumption.

Sources: ***2006:*** *Energy Information Administration (EIA). International Energy Annual 2006 (June-December 2008), web site www.eia.doe.gov/iea.* ***Projections:*** *EIA, World Energy Projections Plus (2009).*

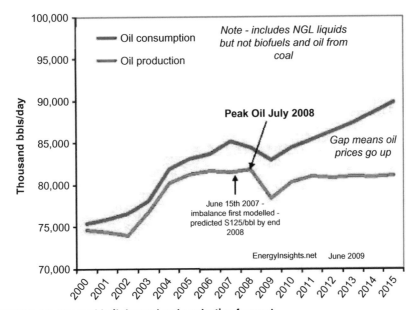

FIGURE 1.10 The world oil demand and production forecast.

In the United States, gasoline prices reached an all-time high during the first week of September 2005 in the aftermath of hurricane *Katrina*. The average retail price was nearly $3.04 per gallon. The previous high was $2.38 per gallon in March 1981, which would be $3.03 per gallon after adjustment for inflation. There are

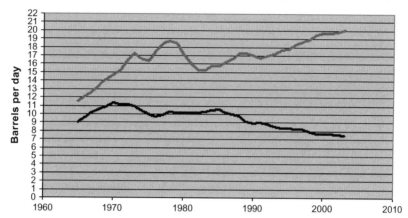

FIGURE 1.11 US oil consumption and production. The USA remains by far the largest oil consumer with declining domestic production now meeting considerably less than 50% of consumption.

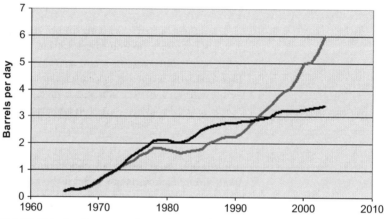

FIGURE 1.12 China's oil consumption and production.

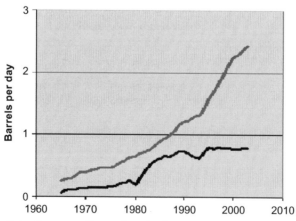

FIGURE 1.13 India's oil consumption and production.

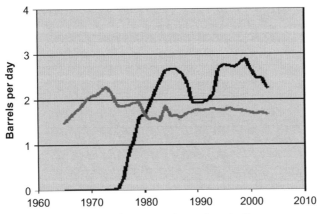

FIGURE 1.14 The United Kingdom's oil consumption and production.

numerous factors that drive the price of a barrel of oil, but the major one is the high world demand for oil. As the demand begins to approach the world's production abilities, the price of a barrel of oil increases.

There are a number of reasons why traders feel that oil supplies might be reduced. One of the most important is turbulence in the Middle East, the world's largest oil-producing region. The war in Iraq, Iran's nuclear program, and internal instability in Saudi Arabia could all lead to a dramatic fall in the supply of oil. Outside the Middle East other oil producers with their own issues, such as the strikes and political problems in Venezuela and potential instability in West Africa, have caused worry for investors.

In late August 2005, *Katrina* crippled the supply flow from offshore rigs on the Gulf Coast, the largest source of oil for the domestic US market. Short-term shut-downs because of power outages knocked out two major onshore pipelines, so at least 10% of the nation's refining capacity was not operating in the wake of the storm. Gas prices in the region, normally $0.70 below the national average, were at $3.12 on August 30 of that year.

World supply came in at 83 million barrels per day during 2004 according to the Department of Energy's EIA calculations. This rate of increase was faster than that of any other date in the past. Despite this, there is increasing discussion about peak oil production and the possibility that the future may see a reduced supply. Even if oil supplies are not reduced, some experts feel the easily accessible sources of light sweet crude are almost exhausted and that in the future the world will depend on more expensive sources of heavy oil and alternatives.

The short-term price of oil is partially controlled by the cartel of the Organization of Petroleum Exporting Countries (OPEC) and the oligopoly of major oil companies. One other important cause is the USD's slump against the Euro. Since oil is traded in dollars, the price must increase for OPEC to maintain purchasing power in Europe.

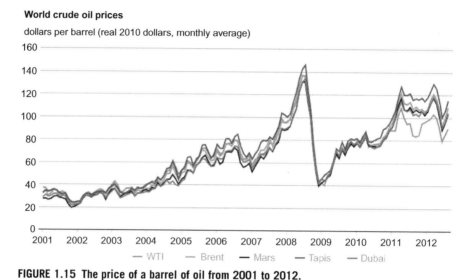

World crude oil prices

dollars per barrel (real 2010 dollars, monthly average)

FIGURE 1.15 **The price of a barrel of oil from 2001 to 2012.**

Sources: Bloomberg, Thomson Reuters. Published by: U.S. Energy Information Administration. Updated: Monthly | Last Updated: 8/31/2012

Figure 1.15 shows the price of a barrel of oil from 2001 to 2012. Many types of crude oil are priced around the world. Variations in quality and location result in price differentials; however, because oil markets are integrated globally, prices tend to move together due to arbitrage. However, now that the world as a whole is becoming more technologically advanced, and the quality of life has increased at a high rate for countries like China and India, the demand is beginning to catch up to the world's production ability. As a result, the cost of a barrel of oil is again climbing to new highs, as shown in the graph in Figure 1.16.

Critics of the oil industry argue that the true cost of oil, and subsquently gasoline, are much higher than wholesale oil markets or retail gasoline prices reflect. Some estimates put the real cost of gasoline near $10.00 per gallon. The hidden oil–gasoline costs consist mainly of tremendous spending on military protection of world oil supplies. The United States alone spends well over $100 billion per year to ensure the free flow of oil from volatile regions of the world.

The argument comes down to this: if such hidden costs were reflected in wholesale and retail prices, instead of being subsidized by the taxpayer, oil and gasoline would be far more expensive than they are today. This hidden cost has the effect of providing oil and gasoline a competitive market advantage over other alternative energy schemes. As the price of a barrel of oil rises, the price of a gallon of gasoline also rises. Figure 1.17 shows the retail gasoline prices for fiscal years 1976 to 2012.

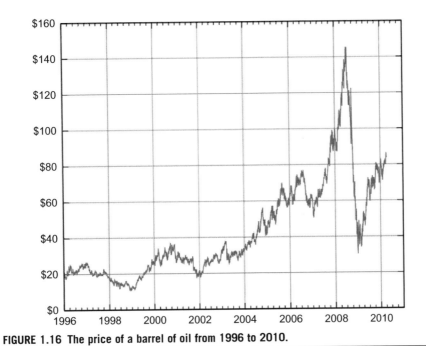

FIGURE 1.16 The price of a barrel of oil from 1996 to 2010.

FIGURE 1.17 The price of a barrel of oil from 1976 to 2012.

1.3.2 Impact of fossil fuel on renewable energy use

Fossil fuel usage causes damage to the environment in two ways: damage caused by the procurement and damage caused by its combustion. The damage is manifested in air pollution, acid rain, and global warming.

Global modernization during the twentieth and twenty-first centuries has created a huge and ever-growing quest for fossil fuels. Today, 90% of US energy comes from the burning of fossil fuels—petroleum, coal, and natural gas. The procurement and use of these energy sources does great harm to the environment, even in the absence of disasters (e.g., oil tanker accidents).

Environmental pollution caused by fossil fuel procurement

Fossil fuels are named as such because they are combustible organic materials derived from the remains of former life; common forms are petroleum, coal, and natural gas. These chemicals are desirable for many reasons, among which are their historical abundance and relative ease of conversion to useful energy via combustion. They also tend to have high energy densities—only one barrel of oil is required to extract, refine, and distribute about 30 barrels. Although much attention has been drawn to the deleterious affects of fossil fuel combustion on the environment, significant damage can be done in extracting and delivering this form of energy.

Damage due to petroleum procurement

About 60 tanker accidents are reported annually in global waters. Accidents like these dump about 280,000 gallons of petroleum in and around US coastal waters per year. However, it is routine tanker operations that causes the majority of spillage into the oceans. Three million tons, or 21 million barrels, of petroleum hydrocarbons are introduced into the ocean annually in the United States alone.

Figure 1.18 shows the devastation that spilled oil can bring to animals. Oil spills kill wildlife by destroying the insulation value of fur and feathers causing hypothermia and death.

FIGURE 1.18 Oil spills kill wildlife.

Other damage from noncombusted petroleum products comes from leakage of underground storage tanks. The Environmental Protection Agency (EPA) conservatively estimates that 50,000 out of 1,000,000 underground tanks are presently leaking.

Damage due to natural gas procurement

Natural gas extraction and transportation infrastructure can be installed on farms, in forests, and on public lands and parks. The delivery infrastructure cuts through ecosystems displacing plants and animals. When piping is laid in and/or under lake and ocean floors, the installation process can stir up harmful sediment.

Explosions of gas-carrying vessels could cause a severe toll on the environment and on human lives. Presently, some groups in southern Massachusetts and Rhode Island are fighting a proposal to site a huge liquid natural gas (LNG) facility in Fall River—only feet away from a residential neighborhood. To access this site, LNG tankers would pass by populated shorelines, such as Providence harbor, where an explosion of tankers could kill 3000 people instantly and severely burn 10,000 more. Environmental damage to waterways, woodland areas, and animals would likely be severe as well.

Damage due to coal procurement

Coal excavation damages land areas; techniques, such as "long wall" mining, can cause the land to sag. This technique dispenses with any attempt to reinforce the land, as mining equipment cuts away large swaths of material underground. When the mined material is removed, surface land (often cropland) can droop 5 to 6 feet. Roads, homes, and utility connections can buckle and trees die, wreaking havoc on farm drainage and wells.

Strip mining is also harsh on the land, leaving it barren for future use unless special topsoils are used. Water passing through mines can pick up pyrite, a sulfur compound in the coal, causing a dilute acid to form, which subsequently runs into nearby rivers and streams.

Miners face daunting occupational hazards. Explosive substances (e.g., methane and coal particulate) are present in mines. If ignited, these materials could kill miners instantly or suffocate them with carbon monoxide fumes. Long-term risks, such as dust inhalation, are present as well (see Figure 1.19). In most cases of black lung disease, small stains and hadened areas of swelling develop as a reaction to the dust. Coal malules and nodules, a small collection of dust and scarring, are distributed throughout the lungs.

Environmental pollution caused by fossil fuel combustion

Conventional energy sources can cause several different types of pollution. Some of the most common ones are air pollution, acid rain, and greenhouse gasses. As a result of fossil fuel combustion, chemicals and particulates are released into the atmosphere. Common examples include carbon monoxide, carbon dioxide, hydrocarbon, nitrogen oxide, and sulfer dioxide.

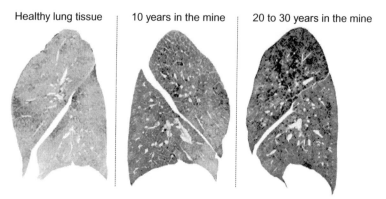

Healthy lung tissue   10 years in the mine   20 to 30 years in the mine

FIGURE 1.19 Images of a coal miner's lung.

Source: Courier-Journal archives

Carbon monoxide (CO) is a product of incomplete combustion and occurs when carbon in the fuel is partially oxidized rather than fully oxidized to carbon dioxide.

Carbon dioxide (CO_2) is a "greenhouse gas" that traps the Earth's heat and contributes to the potential for global warming.

Hydrocarbon emissions result when fuel molecules in the engine do not burn or burn only partially. Hydrocarbons react in the presence of nitrogen oxides and sunlight to form ground-level ozone, a major component of smog.

Nitrogen oxide (NO_x), when under the high-pressure and temperature conditions in an engine, nitrogen and oxygen atoms in the air react to form various nitrogen oxides, collectively known as NO_x. Nitrogen oxides, like hydrocarbons, are precursors to the formation of ozone. They also contribute to the formation of acid rain.

Sulfur dioxide (SO_2) contributes to acid rain.

Conventional energy sources cause pollution

Coal is expected to surpass petroleum and natural gas as the fuel of choice for power stations. However, coal is also the leading cause of smog, acid rain, global warming, and air toxins. In an average year, a typical coal plant generates 3,700,000 tons of CO_2, 10,000 tons of SO_2, and 720 tons of CO.

Automobiles are another culprit in the creation of pollution. Most ground-level ozone pollution is caused by motor vehicles, which account for 72% of NO_x and 52% of reactive hydrocarbons (principal components of smog). A gallon of gasoline weighs just over 6 pounds. When burned, the carbon in it combines with oxygen from the air to produce nearly 20 pounds of CO_2. The different types of pollutions and their effects are explained in detail in the following sections.

Pollutants in the air inside houses and offices are known as indoor air pollution. It can sometimes have more devastating affects in third-world countries than regular air pollution. More than half of the world's population relies on dung, wood, crop waste, or coal to meet their most basic energy needs. Cooking and heating with such solid fuels on open fires or stoves without chimneys leads to indoor air pollution.

Ground ozone pollution should not be confused with the ozone layer found in the troposphere, which helps protect us from the Sun's ultraviolet rays. Ground-level, or "bad" ozone, is not emitted directly into the air, but is created by chemical reactions between oxides of nitrogen and volatile organic compounds (VOC) in the presence of sunlight. Emissions from industrial facilities and electric utilities, motor vehicle exhaust, gasoline vapors, and chemical solvents are some of the major sources of NO_x and VOC, as shown in Figure 1.20.

Effects of pollution

As reliance on conventional power stations and automobiles grows, so does the amount of pollution released in the air. Pollution has several harmful effects on health and the environment.

Health effects

Air pollution can have several detrimental heath effects on many organisms including humans. Many problems have been recorded due to pollution; the following are only a few of them:

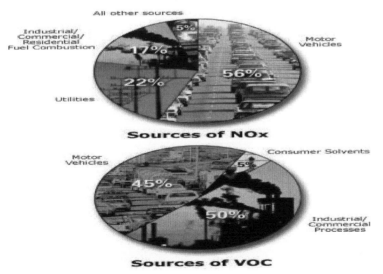

FIGURE 1.20 Sources of ground-level ozone pollutants.

- Acute respiratory problems—temporary decrease in lung capacity and inflammation of lung tissue
- Impairment the body's immune system
- Premature deaths caused by power plant pollution are increasing
- Birth defects, low birth weight, and infant deaths
- Increase in a person's risk of cancer-related death.
- Healthy, active children are 3 to 4 times more likely to develop asthma.
- Ground-level ozone can cause respiratory disease, cardiovascular disease, throat inflammation, chest pain, and congestion.
- Carbon monoxide reduces the flow of oxygen in the bloodstream and is particularly dangerous to persons with heart disease.

According to the World Health Organization (WHO), every year indoor air pollution is responsible for the death of 1.6 million people—that's one death every 20 seconds.

Conventional energy sources have many disadvantages. The most telling evidence is pollution and its affect on health and the environment. Table 1.1 gives a summary of power sources and their effects. Steps taken by government and consumers alike have resulted in improvements in air quality. However, advances will need to be made using the latest technology in renewable energy and alternative energy source vehicles to continue this improvement.

The impact of desposition

When NO_x and SO_2 combine with water in rain clouds, they form nitric and sulfuric acids. The acids pour down on the land and poison lakes and streams, making them too acidic for plant and animal life. The general population is largely unaware of acid rain despite its devastating effects on the environment. As a result, acid rain cannot claim to have a current, significant impact on renewable energy sources and how they are used. However, recent events have triggered a sudden concern over the use of petroleum, its availability, cost, and our reliance on it. Although primarily economically driven, this surge of common concern, combined with renewed education on global warming and pollution in general, has given environmentalists

Table 1.1 Fossil Fuel Burning Products and Their Impact on the Environment

Source	Energy used	Pollutants released	Health effects	Environmental effects
Power stations	Coal, gas	CO_2, SO_2, CO	Cancer-related deaths,	Greenhouse effect, smog, acid rain
Automobiles	Petroleum	Hydrocarbon, NO_x	Respiratory disease, cardiovascular disease	

the ammunition needed to start a more dramatic campaign for using renewable energy sources. As a result acid rain can, and should, play a significant role in promoting the use of cleaner energy sources.

Acid rain is only one of two forms of pollution known as "deposition," typically generated by factories, power plants, vehicles, and anything sending pollutants into the air. This particular form is called *wet deposition*, which is integrated into cloud formations. It is carried by the winds and eventually released when clouds produce precipitation (e.g., rain, sleet, hail, snow).

Such pollution is the most devastating because it is impossible to control which geographical areas will be affected by it, as shown in Figure 1.21. Wherever the winds go and wherever clouds produce precipitation, there is a potential for ground contamination by acid rain and wet deposition. As a result, acid rain caused by a factory on the US West coast can end up affecting ground vegetation and groundwater of the central United States, or even farther away.

In contrast, *dry deposition* consists of pollutants that do not get integrated in cloud formations and always fall back to the ground close to the place where produced in the first place.

Acid rain typically consists of the following three chemical compounds monitored by the EPA:

- Sulfate (SO_4)
- Nitrate (NO_2)
- Ammonium (NH_4)

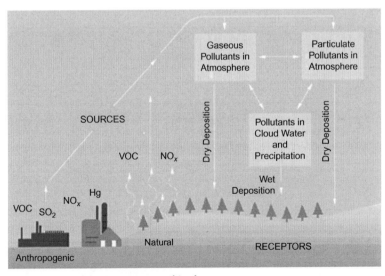

FIGURE 1.21 The process that creates acid rain.

• The compounds can vary drastically, even over the years, and trends are difficult to predict other than that levels everywhere are higher than they should be. Of course, dry deposition can play a part in the levels observed, as other forms of pollution including illegal dumping, commercial dumping, and groundwater contamination due to lack of containment and control, direct wet release into rivers (even when meeting levels as regulated by the EPA)

All areas of the world can be affected by acid rain; however, a smaller subset of those areas can be effectively monitored for contamination directly attributable to it. Those areas must be far enough away from other potential sources of contamination (e.g., dry deposition, illegal dumping, direct release). Some of them are the following:

• Remote forests
• Populated areas far from commercial centers and factories
• Ice shelves

Other areas, almost all not monitored for acid rain's effects, are the various water bodies of the world—rivers, lakes, seas, and oceans. Because so many sources of pollution can contribute to the levels measured in bodies of water, it is difficult to pinpoint the source of the pollution. However, the effects observed cannot be neglected, and acid rain has to be considered as a potential contributor. Some of the most devastating, most observable affects include degradation of coral reefs and decline in underwater life population and variety. Figure 1.22 shows the potential impact of acid rain and other sources of pollution on coral reefs.

The negative impact of pollution on the environment from acid rain is undeniable. The technology to reverse, or at the very least stop progress of the effects of pollution and acid rain, is already available. The problem is social in nature, in that people do not acknowledge the impact of pollution unless it directly affects them, especially when the solution might be inconvenient.

Because acid rain causes deforestation, as can be seen in Figure 1.23, it most probably increases the effects of global warming exponentially. With all this

FIGURE 1.22 A small patch of bleached coral reef.

FIGURE 1.23 Deforestation caused by acid rain.

information in hand, scientists can more easily make a case for cultural change to save the environment, and the general population is more willing to accept that message because it is becoming more convenient or, more appropriately, more cost effective. Recently, it is politically correct to be environmentally friendly because it also makes financial sense.

The most obvious place to start to remedy the pollution problem is the way people and commodities are transported. Electric vehicles have been introduced several times during the last 50 years, always successfully on an experimental basis. Vehicle manufacturers and oil companies, however, have always managed to keep high-power, gas-burning vehicles more popular. Converting both mass and personal transportation from gas-burning to electric would remove all the mobile sources of pollution spread throughout the world and centralize them wherever power is generated, that is, at power plants.

Of course, other such alternative means of transportation can also be used to replace gas-burning engines: hydrogen-, ethanol-, and methanol-based fuel-cell–powered vehicles. Even natural gas powered vehicles are already available; however, natural gas, much like oil, is in limited supply in worldwide and therefore should not be the alternative solution of choice.

Once the sources of pollution have been centralized by removing them from individual vehicles, they can be more easily controlled. Already pollution-generating factories have to follow EPA regulations to prevent more than certain levels from escaping into the atmosphere. Typically, this is done with filtering, which is difficult because it must be customized for the type of pollutants produced by the individual company. However, an alternative to this problem also exists and is available. From an environmental perspective, there are several clean sources of energy already in use throughout the world that could replace existing coal-fired and gas-burning plants.

Combining the replacement of power plants and replacement of vehicle engines can eliminate nearly all sources of pollution related to transportation. But, as pointed out initially, acid rain is not currently a driving factor to push those technologies. Rather the increase in fuel cost is pushing customers to refocus their buying power toward more fuel-efficient vehicles. This in turn is pushing manufacturers to produce more efficient vehicles to meet demand, currently with hybrid vehicles.

Plug-in hybrid vehicles seem to be nearly a reality for the average consumer. True electric plug-in vehicles have come and gone, primarily because of car manufacturers and oil companies, as well as the realities of the economics to a small extent. But their availability is inevitable as demand for them increases. As a result of the economic demands for fuel-efficient vehicles, scientists are now able to more successfully educate the public on both global warming and its causes.

So, in the short term acid rain is having a small impact on the use of renewable energy sources. But in the long term, the impact is immeasurable because it is one of several significant contributions toward changing the average population's perception of the need to use alternative energy sources.

1.4 GLOBAL WARMING AND RENEWABLE ENERGY USE

In today's world, catch phrases, such as "greenhouse effect," "global warming," and "hybrid vehicles" have become part of everyday vocabulary. We have become more and more aware of daily activites and the effect they have on the environment. This is partially due to mainstream media's recent interest in these topics.

This can be recognized by the recent success of the movie *An Inconvenient Truth,* produced by former Vice President Al Gore, about global warming. It won an Academy Award in 2007. Politicians as well as the US government have also taken steps, such as the Clean Air Act, to reduce pollution. However, the amount of pollution released every year continues to rise. Figure 1.24 shows the dynamics of the greenhouse effect phenomenom.

Global warming is the increase in the average temperature of the Earth's atmosphere and its oceans. The greenhouse effect causes the atmosphere to absorb the Sun's longer wavelengths, thus warming it up. It gets it name from the way glass

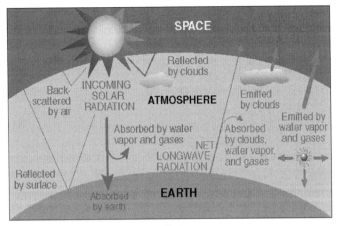

FIGURE 1.24 Illustration of the greenhouse effect.

stores energy and heats a greenhouse. This storage of heat causes the temperature inside to rise. The greenhouse effect is one specific force causing global warming.

Greenhouse gases (GHGs) are components of the atmosphere that contribute to the greenhouse effect. Some GHGs occur naturally in the atmosphere, while others result from human activities such as burning of fossil fuel. Greenhouse gases include water vapor, carbon dioxide, methane, nitrous oxide, and ozone. As shown by Figure 1.25, power stations and automobiles account for more than 35% of annual

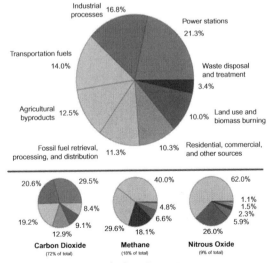

FIGURE 1.25 Annual greenhouse gas emissions by sector.

GHG emissions. Greenhouse gases are made of water vapor, carbon dioxide, and methane, which are all natural gases

The Earth's climate has changed throughout history, from glacial periods (or "ice ages)" when ice covered significant portions to interglacial periods when ice retreated to the poles or melted entirely—that is, the climate has changed continuously. Scientists have been able to piece together a picture of the Earth's climate dating back from decades to millions of years ago by analyzing a number of surrogate, or "proxy," climate measures (e.g., ice cores, boreholes, tree rings, glacier lengths, pollen remains, and ocean sediments) and by studying changes in the Earth's orbit around the Sun. Since the Industrial Revolution ($\sim$1750), human activities have substantially added to the amount of heat-trapping GHSs in the atmosphere. The burning of fossil fuels has also resulted in emissions of aerosols that absorb and emit heat and reflect light.

Figure 1.26 shows the total amount of carbon emissions due to the burning of fossil fuels for the world's energy needs. Conventional energy sources, such as thermal power stations and automobiles, are the leading cause of global warming. Thermal power stations, which contribute up to 69% of the world's energy, use oil, gas, and coal as fuel. However, these three sources combine for the greatest CO_2 emmision in the world.

Greenhouse gas concentrations in the atmosphere will increase during the next century unless CO_2 emissions decrease substantially from present levels (Figures 1.27 through 1.29). Since the start of the Industrial Revolution there have been profound effects on the Earth's atmosphere from the release of increasingly high levels of greenhouse gases into the air. Figure 1.30 shows the projected world carbon dioxide (CO_2) emissions by country.

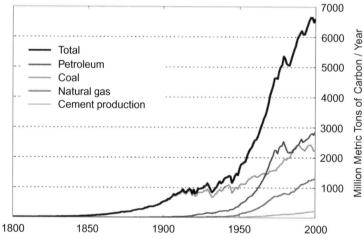

FIGURE 1.26 Global fossil carbon emissions.

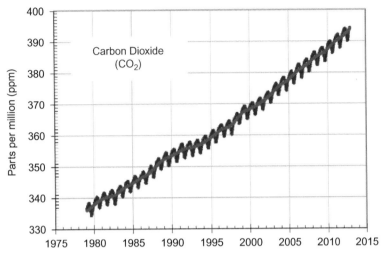

FIGURE 1.27 **The concentration of carbon dioxide.**

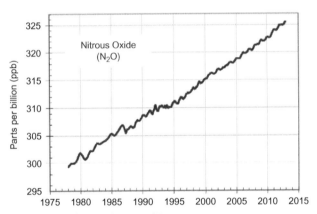

FIGURE 1.28 **The concentration of nitrous oxide.**

1.4.1 **Global warming in general**

Increased CO_2 and other GHG concentrations are likely to raise the Earth's average temperature a considerable amount. The average surface temperature of the Earth is likely to increase by 2.5 to 10.4 °F (1.4–5.8 °C) by the end of the twenty-first century, relative to 1990 (see Figure 1.31). This projected rate of warming is about 2 to 10 times greater than the warming observed during the twentieth century and may represent a warming rate unprecedented for at least the last 10,000 years.

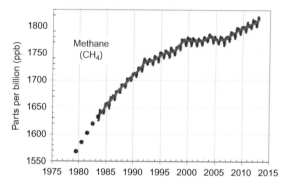

FIGURE 1.29 The concentration of methane.

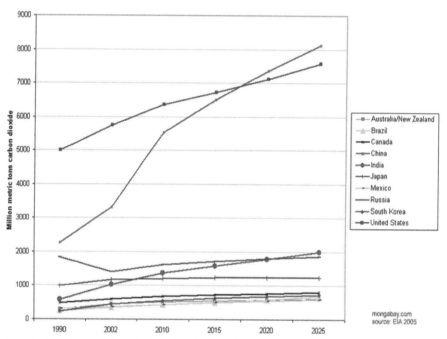

FIGURE 1.30 Worldwide emission of carbon dioxide by country from 1990 to 2005.

Warming will not be evenly distributed around the globe; land areas will warm more than oceans in part due to the water's ability to store heat. High latitudes will warm more than low latitudes partially because of positive feedback effects from melting ice. The northernmost regions of North America, and northern and central Asia, could warm substantially more than the global average. In contrast, projections

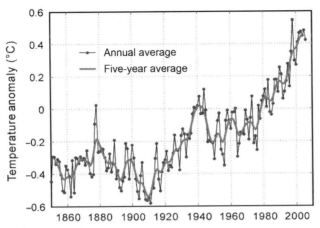

FIGURE 1.31 Global temperature change.

suggest that the warming will be less than the global average in South Asia and southern South America. The warming will differ by season with winters being warmer than summers.

Global warming from the greenhouse effect could also raise sea levels in the next century and even more in the future. Mountain glaciers and ice sheets melting into the oceans are two main components that cause sea levels to rise. James G. Titus states in his Land Use Policy of 1990: "Rising sea levels would inundate deltas, coral atoll islands, and other coastal lowlands, erode beaches, exacerbate coastal flooding, and threaten water quality in estuaries and aquifers."

1.4.2 Global warming and sea levels

Figure 1.32 shows projections of sea-level rise from 1990 to 2100, based on an Inter-governmental Panel on Climate Change (IPCC) temperature projection for three different GHG emissions scenarios (pastel areas, labeled on right). The gray area represents additional uncertainty in the projections due to uncertainty in the fit between increases in temperature and sea levels. All these projections are considerably larger than the sea-level rise estimates for 2100 provided in IPCC AR4 (pastel vertical bars), which did not account for potential changes in ice sheet dynamics and are considered conservative. Also shown are the observations of annual global sea-level rise over the past half century (red line) relative to 1990.

Rising sea levels will cause extreme floods and ruin water quality, which will ultimately cause a wave of extinction among many marine animals. Increasing GHGs could also cause more frequent severe hurricanes, snow and ice storms, tornadoes, and tsunamis, which have escalated in a number during recent years.

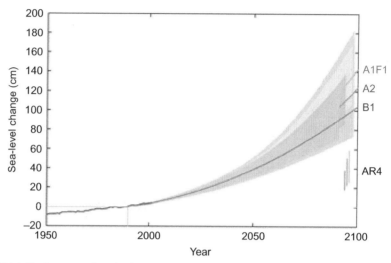

FIGURE 1.32 Sea-level rises projection.

1.4.3 Other global warming concerns and directions

Global warming is unpredictable and can cause droughts in some areas and harsh flooding in others. The greenhouse effect has caused damage all over the world. The most significantly discussed in the media is the bleaching of corals. There are many causes that have led to this, including fertilizers and household products entering and polluting the reef and overfishing. However, growing sea temperatures are responsible for most cases of coral bleaching. The pictures in Figure 1.33 show the drastic effect that bleaching has had on a coral reef. The first image is what a healthy reef looks like, and the second shows what the reef looks like after the color and life is taken away from this natural wonder.

FIGURE 1.33 Coral reef bleaching.

In 1997, the Kyoto Treaty was released in response to growing temperatures in the atmosphere. The Treaty agreed to curb the use of greenhouse gases. Sadly, none of the major industrial countries signed the agreement, arguing that science has not yet proved that humans are the cause. Hopefully, after the IPCC's news release in February 2007, these countries will reconsider and we can start saving our planet.

According to several recent studies, even if the composition of today's atmosphere was fixed, which would imply a dramatic reduction in current emissions, surface air temperatures would continue to warm (by up to 1 °F per year). The studies suggest that a portion of the warming associated with past human activity has not yet been realized because of the heat being stored in the ocean; thus, the Earth is committed to continue to ger warmer. In addition, many of the GHGs that have already been emitted will remain in the atmosphere for decades or longer and will continue to contribute to warming.

With the results of recent studies on global warming and its causes and impacts on our environment, state and federal government agencies are beginning to take notice. There is an increasing interest in what can be done to slow or prevent global warming. However, because of a lack of infrastructure for energy sources (e.g., renewables) and the political influence of current energy company policies, real changes in how we produce energy are slow, but there is hope. Some states have begun to increase their reliance on renewable energy by enacting supportive policies;by reducing barriers to the adoption of renewable technologies; and by encouraging individual, business, and government purchasers of energy to use renewable sources and electric vehicles.

Using energy more efficiently and moving to renewable energy (e.g., wind, solar, geothermal, and biomass) would significantly reduce emission of heat-trapping gases. The United States currently produces 70% of its electricity from fossil fuels, such as coal, natural gas, and oil, but only 2% from renewable sources. Since the burning of fossil fuels releases large amounts of CO_2—the leading cause of global warming—but renewable energy does not, increasing the share of electricity generated using renewable resources is one of the most effective ways to reduce global warming emissions.

1.4.4 What can be done to curb global warming?

There have been a lot of suggestions on what can be done to help reduce the global warming poblem. The solution must include a way in which we can cut the emission of CO^2 and also try to use the most efficient source of energy.

Renewable energy

The burning of fossils for generating electricity contributed a massive 40% of the carbon dioxide emission in 2002. A way to reduce this pollution is to find alternative energy sources. Harnessing the power of the wind or the Sun could be very helpful.

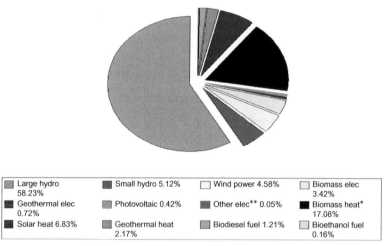

■ Large hydro 58.23%	■ Small hydro 5.12%	□ Wind power 4.58%	■ Biomass elec 3.42%
■ Geothermal elec 0.72%	▨ Photovoltaic 0.42%	▨ Other elec** 0.05%	■ Biomass heat* 17.08%
■ Solar heat 6.83%	■ Geothermal heat 2.17%	▨ Biodiesel fuel 1.21%	□ Bioethanol fuel 0.16%

FIGURE 1.34 The world's renewable energy for 2005.

These are energies that are readily available and create no pollution. Figure 1.34 shows the world's renewable energy production for the year 2005.

Reduce emissions by using alternative-power vehicles

Cars, buses, and trucks are responsible for about 20% of the CO^2 emissions in the United States. This is so because most US vehicles are powered by the internal combustion engine (ICE), which burns gasoline. The less the gas mileage of the vehicle is, the more contribution it makes to the emission of CO^2. This problem can be rectified by finding a clean-burning fuel that produces little to no carbon dioxe

1.5 CONCLUSION

Rising oil prices, combined with global warming, acid rain, and pollution caused by the conventional sources of energy, will definitely accelerate the world's interst in reneable enegy sources. If natural resources continue to be exploited at the current staggering rate, there will be nothing left to exploit.

Conservation strategies have been implemented, especially at times when an energy shortage has occurred, but they are all but forgotten when the shortage passes. Unfortunately, conserving resources at this point will only temporarily postpone the inevitable—total energy resource exhaustion. Unless the lifestyles and transportation habits of the ever-increasing population, which demands more and more resources, are dramatically changed, or new energy sources are discovered, conserving won't be enough.

References

[1] How stuff works, http://auto.howstuffworks.com/fuel-efficiency/fuel-economy/hydrogen-economy1.htm; 2012 [accessed 10.01.12].

[2] http://upload.wikimedia.org/wikipedia/commons/b/bb/Major_greenhouse_gas_trends.png.

[3] Air Now, US air-quality summary, http://airnow.gov/index.cfm?action=goodup.page1; [accessed 20.02.12].

[4] Holcomb Station Expansion Project. Power plant major components, www.holcombstation.coop/Technology/Components.cfm; 2010 [accessed 12.05.10].

[5] Wikipedia. Francis turbine, http://en.wikipedia.org/wiki/Francis_turbine; 2010 [accessed 12.05.10].

[6] Kandaka.com. US electric power generation by fuel type, http://bannaga.files.wordpress.com/2007/02/figes1.gif; 2012 [accessed 20.9.12].

[7] US Department of Energy/EIA. Annual energy outlook 2007, ftp://tonto.eia.doe.gov/forecasting/0383(2007).pdf; 2012 [accessed 28.09.12].

[8] Wikia. Renewable energy, http://sca21.wikia.com/wiki/Renewable_energy; [accessed 20.09.12].

[9] US Department of Energy/EIA. World energy projection plus, www.eia.doe.gov/iea; 2012 [accessed 20.09.12].

[10] US Energy Information Administration. Electricity demand and supply, www.eia.doe.gov/oiaf/aeo/pdf/trend_3.pdf; 2010 [accessed 12.05.10].

[11] US Energy Information Administration. Electricity infocard 2006, www.eia.doe.gov/neic/brochure/elecinfocard.html; 2010 [accessed 12.05.10].

[12] US Energy Information Administration. Electricity, www.eia.doe.gov/oiaf/ieo/pdf/electricity.pdf; 2010 [accessed 12.05.10].

[13] EIA and Bureau of labor statistics, www.gasoilprices.info/brent.html; [accessed 10.01.12].

[14] Gregor. US global oil production update and EIA data changes, http://gregor.us/eia/global-oil-production-update-and-eia-data-changes/.

[15] EIA. International energy annual, 2006 (June–December 2008), www.fypower.org/pdf/EIA_IntlEnergyOutlook(2006).pdf; [accessed 10.01.12].

[16] Wikipedia. Petrolium, http://en.wikipedia.org/wiki/World_oil_consumption; 2010 [accessed 14.05.10].

[17] Oil spill kills wildlife, Google image, http://hot1079philly.com/tag/oil-spill-kills-wildlife/; 2012 [accessed 9.28.12].

[18] Ecosystem. Oil production and consumption for certain countries, www.hubbertpeak.com/nations/2004/; 2010 [accessed 14.05.10].

[19] NOAA. Natural and anthropogenic stressors on Palau's Coral Reefs, http://coris.noaa.gov/about/eco_essays/palau/stressors.htmll; 2010 [accessed 28.09.12].

[20] Wikipedia. Pollution, http://en.wikipedia.org/wiki/Pollution; 2010 [accessed 14.05.10].

[21] US Environmental Protection Agency. General information, www.epa.gov/; 2010 [accessed 14.05.10].

[22] Trees killed by acid rain in the Great Smoky Mountains, www.scienceclarified.com/A-Al/Acid-Rain.html; 2010 [accessed 20.05.10].

[23] US Environmental Protection Agency. Climate change science, www.epa.gov/climatechange/science/futurecc.html; 2010 [accessed 20.05.10].

[24] Coral reef pictures, www.abc.net.au/reslib/200406/r23761_58708.jpg; 2010 [accessed 20.05.10].

[25] Titus J. Greenhouse effect, sea level rise, and land use. Land Use Pol April 1990; 7(2):138–53, 3 Jan 2007.

[26] Publication of the National Oceanic and Atmospheric Administration (NOAA). 28 Feb 2007, NOAA Office of Communications, www.coral.noaa.gov/cleo/coral_bleaching.shtml; 2010 [accessed 14.05.10].

[27] Wikipedia. Global warming, http://en.wikipedia.org/wiki/Global_warming; 2010 [accessed 14.05.10].

[28] Gas Imaging Technology, LLC, www.gitint.com/sherlock_env2.html; 2012 [accessed 28.09.12].

[29] US EPA. Acid rain, www.epa.gov/acidrain/what/index.html; 2012 [accessed 28.09.12].

[30] The National Academies Press, www.nap.edu/openbook.php?record_id=12782&page=244, year?, (accessed?).

[31] Gas and oil prices, www.gasoilprices.info/brent.html; 2012 [accessed 28.09.12].

[32] Tag Archive; Black lung, http://parentheticalhilarity.files.wordpress.com/2010/11/black-lung.jpg; 2012 [accessed 28.09.12].

[33] Online magazine of the American Enterprise Institute, www.american.com/archive/2012/march/why-are-gasoline-prices-high-and-what-can-be-done-about-it; 2012 [accessed 28.09.12].

[34] Energy Insights. Global oil production, www.energyinsights.net/cgi-script/csArticles/articles/000000/000085.htm; [accessed 10.01.12].

[35] Bloomberg, Thomson Reuters. U.S. Energy Information Administration. Updated: Monthly | Last Updated: 8/31/2012.

Photovoltaic

2.1 INTRODUCTION

Photovoltaic cells are devices that convert light into electricity. Because the source of light (or radiation) is usually the Sun, they are often referred to as *solar cells*. The word *photovoltaic* comes from "photo," meaning light, and "voltaic" for voltage. The output of a photovoltaic cell is direct current (DC) electricity.

The direct energy conversion of light to electricity was first reported in 1839 by a Frenchman named Becquerel, who observed a difference in electrical potential between two electrodes immersed in electrolyte. The potential varied with light intensity. Although there have been many significant historical developments between 1839 and the present, most discussions of photovoltaic and the modern silicone cell dates back to 1954 and the work done at Bell Labs.

In 1960, photovoltaic cells were essentially handmade, cost approximately $1000 per peak watt of power and were used almost exclusively for space applications. In 1970, cells were manufactured in batch processes, cost approximately $100 per peak watt, were primarily used in space applications, but were making an entry into stand-alone applications. In 1980, larger quantities of cells were produced in larger batch processes, cost approximately $10 per peak watt, saw significant terrestrial applications (primarily stand-alone, remote devices), and began to be used for grid-connected residential applications.

Today, we are on the threshold of continuous-process manufacturing and photovoltaic modules are beginning to be purchased (in large orders) for approximately $4 per peak watt. In the United States, there are several thousand (mostly very small) residential systems and a few central power stations in operation or under development—primarily in California. In addition, there is a rapidly growing market for photovoltaic consumer products such as watches, calculators, radios, televisions, battery rechargers, and so on.

As our conventional source of energy—fossil fuel—rapidly depletes, the search for alternative energy sources has become an integral issue in modern industrialized society. Unavoidably, the world's energy demands and cost will rise. To keep pace with these demands, unlimited energy sources must be researched and optimized. Among the most infinitely available and clean source is solar energy.

Solar energy can be harvested with photovoltaic (PV) cells—solar cells. When a cell is exposed to solar radiation, a voltage potential builds up across it. If a load is connected to the cell, this voltage results in a current flow through the load. A typical solar cell produces approximately 0.5 volts (V) and a current that significantly depends on the intensity of the sunlight and the area of the cell. To get more usable values of voltage and current, solar cells are connected in series to increase voltage, and the series of cells are connected in parallel to increase current output. Typically, 20 to 60, or more, cells are packaged together with a transparent cover (usually glass) and a watertight seal to form a module. In turn, these modules are wired together in a series–parallel combination to form a panel that best meets the needs of the application. An array is a group of panels that have been connected mechanically and electrically.

Photovoltaic systems are used today in many applications such as battery charging, water pumping, home power supply, satellite power systems, and so forth. Although they have the advantage of providing an unlimited and clean source of energy, their installation cost is high and still have relatively low conversion efficiency. Commercially available PV cells have average efficiencies of 11% and higher efficiencies are only around 15%.

2.2 SOLAR RADIATION CHARACTERISTICS

2.2.1 Solar constant

Although artificial light can be used to power photovoltaic devices, their value lies in their ability to use free and renewable sunshine. A photovoltaic device outside the Earth's atmosphere that maintains normal incidence to the Sun's rays receives an approximately constant rate of energy. This amount, called the solar constant, is 1.353 kW/m^2 (428 Btu/hr*ft^2). The solar constant and its associated solar radiation spectrum immediately outside the atmosphere are determined solely by the nature of the radiating source (i.e., the Sun) and the distance between the Earth and the Sun.

2.2.2 Hourly and seasonal variations in solar radiance

Terrestrial applications of photovoltaic (or any solar) devices are complicated by two variables: (1) the Earth's rotation about its axis and revolution about the Sun and (2) atmospheric effects. The rotation produces hourly variations in power intensities at a given location on the ground and completely shades the device during nighttime hours. In addition, a device located in the Northern Hemisphere receives more energy during summer than in the winter, thus giving rise to seasonal variations in power intensities.

2.2.3 Direct and diffuse sunlight

The presence of the atmosphere and associated climatic effects both attenuate and change the nature of the solar energy resource. The combination of reflection, absorption (filtering), refraction, and scattering results in highly dynamic radiation

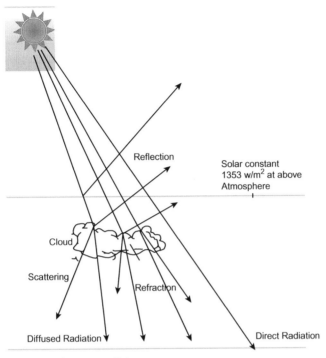

FIGURE 2.1 Direct and diffused sunlight.

levels at any given location on the Earth's surface. Because of cloud cover and the scattering of sunlight, the radiation received at a specific point is made up of both direct (or beam) sunlight and diffuse (or scattered) sunlight. This distinction is important because concentrating devices rely on their ability to focus direct sunlight. (See Figure 2.1.)

There is a considerable regional variation in the amount of direct radiation received. In the United States, the desert Southwest receives the most. Consequently, many of the initial applications of concentrating and tracking flat-plate devices have been located in California, Arizona, and New Mexico.

2.2.4 Air mass

Air mass, defined as $1/\cos\theta$, where θ is the angle between the sun beam and a line normal to the surface of the solar cell, is a useful quantity to deal with atmospheric effects. It indicates the relative distance that light must travel through the atmosphere to a given location. Because there are no effects due to air attenuation immediately outside the Earth's atmosphere, this condition is referred to air mass zero (AM0). Air mass one (AM1) corresponds to the Sun being directly overhead. Air mass 1.5 (AM1.5) is considered more representative of terrestrial conditions and is commonly

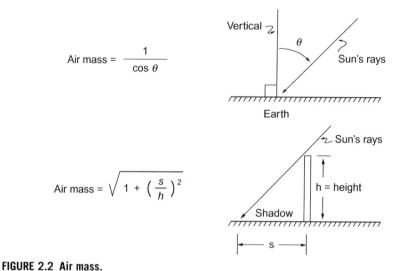

FIGURE 2.2 Air mass.

used as the reference condition in rating photovoltaic modules and arrays. The spectral distribution of sunlight can be plotted for various air mass values, thus providing a useful indication of atmospheric attenuation effects. (See Figure 2.2.)

Irradiance

Power is the rate of doing work in units: kW, or watts. Irradiance is radiant power per unit area.

- Units: kW/m^2, or $watts/m^2$, or mW/cm^2
- Measurement: Pyranometer (e.g., Eppley PSP)
- Peak value: $1\ kW/m^2$ ($1000\ watts/m^2$, $100\ mW/cm^2$)

Insolation

Energy is the capacity for doing work in units: kWh. Insolation is radiant energy per unit area. Irradiance is summed over time—units: kWh/m^2.

2.2.5 Peak sun hours

The amount of terrestrial power received on a properly tilted surface on a clear day is approximately $1\ kW/m^2$, which is referred to as the peak (or full) sun condition. Note that it is considerably less than the solar constant ($1.353\ kW/m^2$). This is indicative of the attenuating nature of the atmosphere. If we have $1\ kW/m^2$ of sunlight for one hour, we receive $1\ kWh/m^2$ or one (peak) sun hour of energy. For a typical day in the US Sunbelt, we receive about 4.5 to $6\ kWh/m^2$ per day, which is equivalent to saying we receive 4.5 to 6 peak sun hours per day.

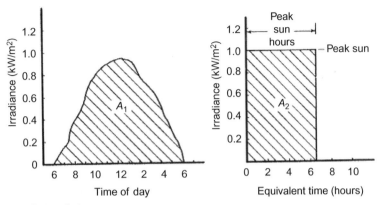

FIGURE 2.3 Solar windows.

The number of peak sun hours per day at a given location is defined as the equivalent time in hours at peak sun conditions (i.e., 1 kW/m^2) that yields the same total isolation. This means that in Figure 2.3 area $A_1 = $ area A_2.

The solar window

The space between the shortest sun trajectotory (shortest day of the year, usually December 21) and the longest sun trajectory (longest day of the year, usually June 21) is called the solar window (Figure 2.4). For a location, it is a measure of how sunny the location is; the wider the solar window, the sunnier the location. (See Figure 2.5.)

Atmospheric effects on solar energy reaching the earth

The solar cell on Earth receives maximum radiation when its surface is perpendicular to the sunbeams. Due to the fact that the Sun's trajectory in the sky varies seasonally, the tilt angle of the collector where the cell lies should also vary if the solar cell is to extract maximum power from the Sun. This can be achieved using a sun-tracking

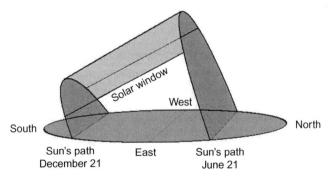

FIGURE 2.4 Solar window.

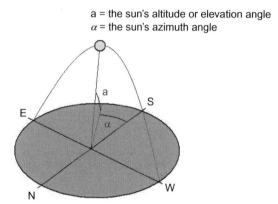

a = the sun's altitude or elevation angle
α = the sun's azimuth angle

FIGURE 2.5 Sun's location in the sky.

technique. That is why the collector should have almost a zero tilt angle at noon in June and much higher angle in December at noon. If the tilt angle is fixed, it should be equal to the latitude of the location (Figure 2.6).

Distribution of energy in sunlight

Light is composed of tiny bundles of energy, which act like individual bullets, that travel at an extremely high speed. These bullets are called *photons*. A beam of sunlight is a stream of photons. These photons travel in waves; the shorter the wave length, the higher the speed and the more energy it has. Figure 2.7 shows the difference between the energy level at AM0 and AM1.5, which reflects the air attenuation. It also shows a detailed distribution of the solar spectrum.

A close look at the solar spectrum in the figure reveals the importance of the ozone. It absorbs the ultraviolet photons that travel at a very high speed and will cause a great deal of damage to life on Earth. Most the energy that reaches the Earth's surface lies between 300 to 2500 nm, 50% of this lies between 400 and 700 nm. A great deal of the infrared photons are absorbed by carbon dioxide (CO_2). Water

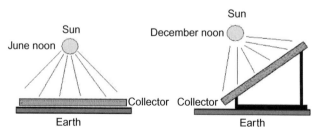

FIGURE 2.6 Collector tilt angle for June and December.

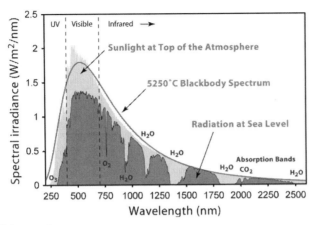

FIGURE 2.7 Solar spectrums.

vapor in the air absorbs 30% of the incident energy. Also, the curve reveals that most of the incident energy occurs at the blue end of the spectrum, which accounts for the color of the sky.

2.3 PHOTOVOLTAIC CONVERSION

To understand how a solar cell converts sunlight into electricity, we have to look at the construction of the solar cell. It is nothing more than a P-N junction diode that has provisions for illumination by the solar spectrum and provides for electrical connections to an external circuit.

An atom is made up of two main parts: first, the sphere you can see in the middle, and second, the electron cloud, which is the random path the electron follows around the atom. The nucleus is made up of *protons,* which have a positive charge, and *neutrons,* which don't have a charge and so are neutral.

Electrons have a negative charge. The total charge of the electrons in regular atoms equals the total charge of the protons; therefore, a regular atom is neutral in charge. The electrons get filled in the orbits according to the formula $2\,N^2$, where N = number of the shell (orbit). The second and third shells are both only allowed to have 8 electrons. (See Figure 2.8.)

2.3.1 Atomic number

The atomic number of an element gives us three pieces of information. First, it gives us the elements' position in the Periodic Table of the Elements. Second, it gives us the number of protons, and, finally, the number of electrons. One of the fundamental concepts in electronics is conduction. Different materials conduct an electrical

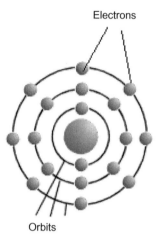

Electrons

Orbits

FIGURE 2.8 Atom construction.

current in many diverse ways. According to their conduction properties, they are classified as follows:

Good conductor: It allows currents to pass through it freely. These elements have 1 to 2 electrons in the last orbit (e.g., metals such as gold or copper).
Bad conductor: It doesn't allow free flow of currents through it; a bad conduction is also known as an insulator. These elements have 6 to 7 electrons in the last orbit (e.g., glass, wood).
Semiconductors: They have conducting properties between the good and bad conductors. These elements have 3 to 5 electrons in the last orbit (e.g., silicon, germanium).

Semiconductors show photovoltaic properties. Semiconducting elements are located in group 14 of the Periodic Table. The element names and symbols are listed here.

Element	Symbol
Carbon	C
Silicon	Si
Germanium	Ge
Tin	Sn
Lead	Pb

Silicon

Silicon is a gray-colored element with a crystalline structure. It is the second most abundant element in the Earth's crust, after oxygen. Silicon is always found in combined form in nature, often with oxygen as quartz and is found in rocks and silica sand.

Germanium

Germanium is another semiconductor found in group 14 of the Periodic Table. It is gray–white with a shimmering white luster.

Doping

Semiconductors can change their electrical properties and degree of conduction by the addition of impurities. The procedure of adding impurities to a semiconductor is called *doping*. The impurities that are used are called *dopants*. There are two different types of dopants: n-type and p-type

N-type dopants

Some of the most common n-type dopants are located in group 15 of the Periodic Table (e.g., phosphorus, P). The n-type dopant adds negatively charged electrons to the semiconductor. Sometimes these dopants are called *donors* because they donate electrons to the semiconductor. Group 15 is shown in the following list.

Element	Symbol
Nitrogen	N
Phosphorus	P
Arsenic	As
Antimony	Sb
Bismuth	Bi

Phosphorus, arsenic, and antimony are donor dopants in silicon, making it n-type. A donor atom is easily ionized, yielding a free (−) electron and leaving behind a positive ion core; see Figure 2.9.

P-type dopants

P-type dopants are found in group 13 of the Periodic Table (e.g., boron, B). These dopants introduce something called *holes* because they steal electrons from the

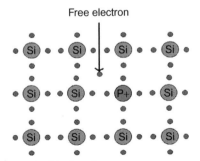

FIGURE 2.9 N-type silicon doped with phosphrous.

semiconductor. As a result of these electron deficiencies, the total negative charge is reduced and the holes can be thought of as positive charges. Group 13 is shown in the following list:

Elements	Symbol
Boron	B
Aluminum	Al
Gallium	Ga
Indium	In
Thallium	Ti

P-type silicon

Boron, aluminum, and gallium are acceptor dopants in silicon, making it p-type. An acceptor atom is readily ionized, yielding a free (+) hole and leaving behind a negative ion core; see Figure 2.10.

Diffusion (of electrons and holes)

Diffusion is a process whereby particles tend to spread out or redistribute as a result of their random thermal motion. Particles migrate from regions of high particle concentration to regions of low particle concentration. (See Figure 2.11.)

P-N junction diode

A *p-n junction* consists of p-type and n-type semiconductor material in intimate contact with one another (with no intervening material of any kind). This is the solar cell. This structure is also called a p-n diode, or simply a diode.

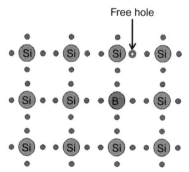

FIGURE 2.10 P-type silicon doped with boron.

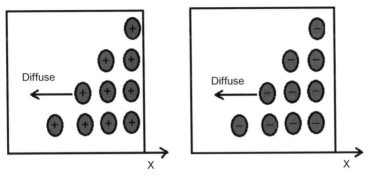

FIGURE 2.11 Diffusion in p- and n-type silicon.

2.3.2 Properties and characterization of a photovoltaic: p-n junction

The charge distribution junction formation created, as shown in the following, gives rise to an electric field and thus a potential difference across the junction.

1. Free electrons in n-region attracted to positive charge in p-region, drift on over
2. Free holes in p-region attracted to negative charge in n-region, drift on over
3. Leaves n-region with net positive charge and p-region with net negative charge, whole crystal neutral
4. Potential barrier formed at junction

This field will soon stop any further diffusion. The potential difference will have an energy called the Energy Gap (E_g); see Figure 2.12. To free an electron from its orbit, this much energy E_g should be given to it.

Solar cell conversion

A solar cell is a p-n junction formed in silicon (Si), gallium arsenide (GaAs), or another material. When solar radiation with proton energies (E_p) that is greater than the E_g strikes a solar cell, electrons in the valence band acquire enough energy to

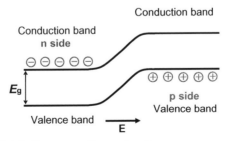

FIGURE 2.12 Charge distribution across the p-n junction.

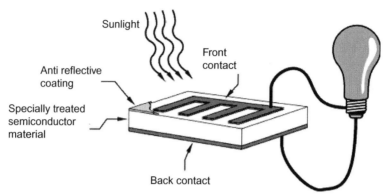

FIGURE 2.13 Principle operation of a photovoltaic cell.

jump to the conduction band, thus producing electron–hole pairs. The electrons travel from the n-side of the junction around the external circuit and back to the p-side of the junction, where they recombine with the holes and close the current path. The current is just a flux of electrons.

If there is no external circuit provided, the n-side would be charged relative to the p-side and a new equilibrium will be reached with a new potential difference called *open-circuit voltage* (V_{oc}). Figure 2.13 illustrates the principle of the operation of a PV cell.

Cell materials

A variety of semiconductor materials are used when fabricating photovoltaic cells and modules. These include single-crystal silicon, polycrystalline silicon, amorphous silicon, and a large number of advanced technology materials—most notably cadmium sulfide and gallium arsenide. Polysilicon is the feedstock for the production of a high-quality, single-crystal silicon sheet.

Limitations on the efficiency of a solar cell

The maximum possible conversion efficiency of regular solar cells is limited to 20 to 30%. This limitation comes mainly as a result of the cell's response to only a portion of the wavelengths available in the solar spectrum. Here are some of the facts:

- Every photon frees only one electron.
- Some light is reflected at the surface of the solar cell and wasted.
- Photons with energies less than the band gap energy pass through the cell without freeing any electrons and are lost.
- Photons with energies larger than band gap energy will have excess energy that is lost too.
- Unused energy will be converted to heat, which will cause the electrons to pass the junction the wrong way, thereby adding to the detrimental influence on conversion efficiency.

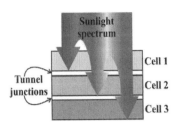

- A stack of single-gap solar cells.
- Each of the cells uses a different part of solar spectrum.
- The open-circuit voltage is the sum of the V_{oc}s of individual cells.
- Requires current matching.

FIGURE 2.14 Multijunction solar cells.

The efficiency is affected by the cell material, which affects band gap energy.

Monocrystalline: 25% efficient
Polycrystalline: 20% efficient
Thin film: 10% efficient

To improve the efficiency multijunction, solar cells are used where each cell uses a different part of the solar spectrum, as shown in Figure 2.14.

2.4 PERFORMANCE EVALUATION OF PV CELLS, MODULES, PANELS, AND ARRAYS

A typical PV cell produces approximately 0.5 V and a current that very much depends on the intensity of the sunlight and the area of the cell. To get more usable values of voltage and current, PV cells are connected in series to increase voltage, and the series of cells are connected in parallel to increase current output. Many cells are packaged together with a transparent cover (usually glass) and a watertight seal to form a *module*. In turn, these modules are wired together in a series–parallel combination to form a *panel* that best meets the needs of the application. An *array* is a group of panels that have been connected mechanically and electrically, as shown in Figure 2.15.

2.4.1 Current–voltage characteristics

The most important performance descriptor for a photovoltaic device (i.e., cell, module, panel, and array) is its current–voltage (*I–V*) characteristic. For a given irradiance, the operating voltage and current vary with load. Current will vary from zero (corresponding to infinite load or open circuit) to a maximum called I_{sc} (corresponding to zero load or short circuit). The maximum operating voltage (V_{oc}) corresponds to open-circuit conditions and the voltage is zero at the short-circuit condition. Thus,

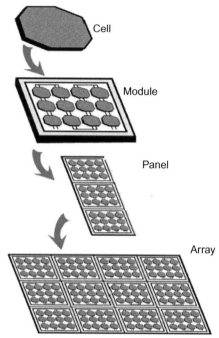

FIGURE 2.15 Photovoltaic array nomenclature.

I_{sc}, and V_{oc} are two limiting parameters that are used to characterize a photovoltaic device for given irradiance and operating temperature.

To evaluate the performance of a solar module, which is usually available in the marketplace, the setup shown in Figure 2.16 is used. It consists of a variable resistive load, an ammeter, and a voltmeter. Figure 2.17 shows the *I–V* characteristics of a PV

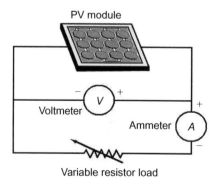

FIGURE 2.16 A set ups for testing a PV module.

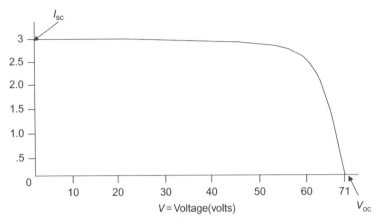

FIGURE 2.17 *I–V* characteristic of a PV mobile module.

mobile module; the shape of the *I–V* characteristics of all PV devices is the same. I_{sc} current and V_{oc} voltage connect through many intermediate points.

2.4.2 Power-voltage characteristics

The power can be calculated for every point on the *I–V* characteristics as $P = V \times I$. It is clear that the power is zero at I_{sc} and V_{oc} points, and it is at maximum somewhere between, $V = 0$ and $V = V_{oc}$, the *P–V* characteristics of the PV mobile module is shown in Figure 2.18.

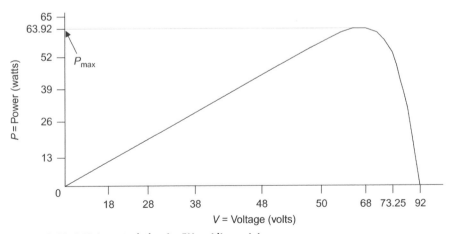

FIGURE 2.18 *I–V* characteristic of a PV mobile module.

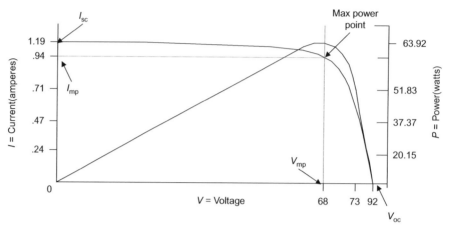

FIGURE 2.19 Maximum power point of a PV mobile module.

2.4.3 **Maximum power point of a PV module**

If we superimpose the *I–V* characteristics curve and the *P–V* characteristics curve, we will identify the point at which the power output of a PV module is at its maximum value, as is clear from the *P–V* curve shown in Figure 2.19. The corresponding point on the *I–V* characteristics curve, where the power is at the maximum, is identified with I_{mp} and V_{mp}.

- I_{mp} is the current at which the power output is at its maximum value.
- V_{mp} is the voltage at which the power output is at its maximum value.

2.4.4 **Effect of resistive load on the operating point**

When a resistive load is connected to a PV module operating point, that is to say the current and the voltage, is decided by intersection of the *I–V* characteristics curve of the PV and the *I–V* charactersitics of the load curve, $V = IR$ (Figure 2.20). The slope of this curve is $1/R$ and the operating point will depend on the value of R. Each slope represents a certain resistance.

Effect of irradiance on cell I–V characteristics

The *I–V* curve of a solar cell is valid for only one irradiance. Every level of irradiance corresponds to one curve. As irradiance increases, the I_{sc} increases linearly and the V_{oc} decreases very slightly, dictating a new curve. Figure 2.21 shows the change in the *I–V* characteristics for various irradiances. Notice that the maximum power points that correspond to different irradiances lie on a line that connects the maximum power points for different irradiances.

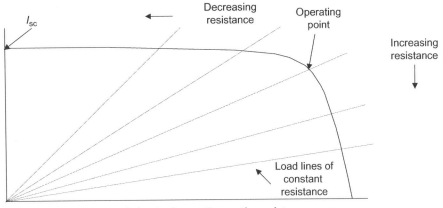

FIGURE 2.20 Effect of resistive load on cell operating point.

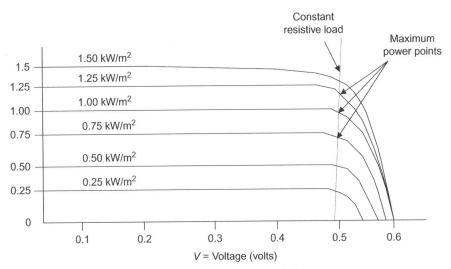

FIGURE 2.21 Effect of irradiance on PV cell *I–V* characteristics.

Effect of irradiance on a PV cell voltage and current

As the irradiance (the light intensity) increases, the I_{sc} increases linearly, but the open-circuit voltage increases logarithmically. (See Figure 2.22.)

Effect of temperature on PV I–V characteristics

As the temperature increases, the open-circuit voltage, V_{oc}, of the PV cell decreases. Also I_{sc} increases very slightly. (See Figure 2.23.)

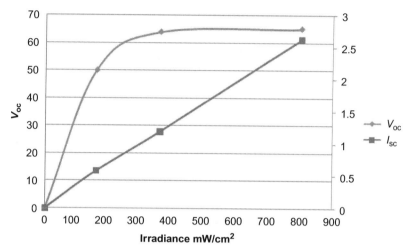

FIGURE 2.22 Effect of irradiance on PV voltage and current (mobile module).

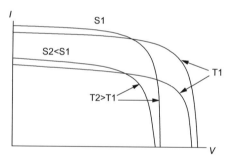

FIGURE 2.23 Effect of cell temperature on PV cell *I–V* characteristics.

Effect of cell temperature on cell short-circuit current, open circuit voltage, and maximum power

In general, as the temperature increases, I_{sc} increases slightly, V_{oc} decreases a lot and the maximum power P_{max} decreases with less percentage than V_{oc}. The effect of temperature is shown in Figure 2.24.

Module peak rating conditions

The module rating is usually given at certain standard conditions:

Irradiance $= 1000 \text{ W/m}^2 =$ One sun
Air mass $= 1.5$
Cell temperature $= 25\ ^\circ\text{C}$

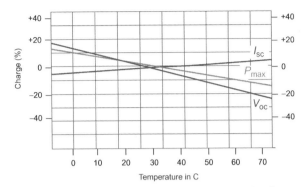

FIGURE 2.24 Effect of temperature on short-circuit current, open-circuit voltage, and maximum power.

Effect of cell temperature on solar cell efficiency

The efficiency of a solar cell is calculated as the power output of the cell, P_{out}, divided by the solar power incident on the cell, P_{in}; where,

$P_{in} = S \times A_c$
$S = $ irradiance
$A_c = $ area of the cell

The efficiency η is given as

$$\eta = P_{out}/P_{in}$$

As the temperature increases, the power output decreases, thus, the efficiency of the solar cell decreases. (See Figure 2.25.) The fill factor (see curve shown in Figure 2.26) equation is as follows:

$$\text{Fill factor} = \text{FF} = \frac{I_{mp}V_{mp}}{I_{sc}V_{oc}} = \frac{P_{max}}{I_{sc}V_{oc}} \tag{2.1}$$

2.4.5 Equivalent circuit of a solar cell

The solar cell can be represented by a circuit that depicts its behavior, and this circuit is called an equivalent circuit (Figure 2.27). The equivalent circuit of a solar cell consists of the current source, I_L, which is a function of the irradiance, or light intensity. Its value is very close to I_{sc}.

- Diode
- Shunt resistance, R_{sh}
- Series resistance, R_s
- Photo generated current, I_L
- Ideal diode current, I_d

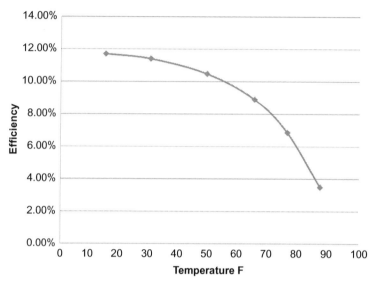

FIGURE 2.25 Effect of temperature on efficiency (mobile module).

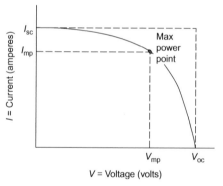

FIGURE 2.26 Fill factor curves.

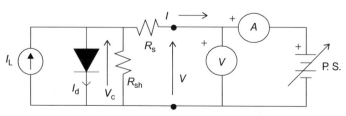

FIGURE 2.27 The equivalent circuit of the solar cell.

The current–voltage characteristics for a solar cell are given by the following equation:

$$I = I_L - I_o(\exp(q(V + IR_s)/AkT) - 1) - (V + IR_s)/R_{sh} \qquad (2.2)$$

where

I_o = the reverse saturation current of the diode
$K_o = q/AkT$
k = Boltzmann's constant
q = charge of an electron
T = absolute temperature
A = ideality factor equal to 1

Ideal solar cell model

The equivalent circuit of an ideal solar cell is shown in Figure 2.28. The series resistance, R_s, is very small and the shunt resistance, R_{sh}, is usually very large; both can be neglected and the result is a simplified or ideal model. This simplified model can be represented by the following equation:

$$I = I_L - I_o[\exp(K_o V_c) - 1] \qquad (2.3)$$

where

I = cell current (A)
I_L = light generated current (A)
I_d = diode current (A)

The diode current is given by this equation:

$$I_d = I_D[\exp(V/V_t - 1)] \qquad (2.4)$$

$$V_t = KT/q \qquad (2.5)$$

where

I_D = the saturation current of the diode
q = electron charge = 1.6×10^{-19} Coulombs
A = curve fitting constant
K = Boltzmann constant = 1.38×10^{-23} J/°K
T = absolute temperature on absolute scale °K

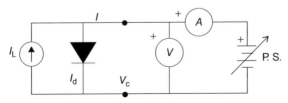

FIGURE 2.28 Simplified equivalent circuit.

The shunt-leakage current is given by the expression. The last term is usually for simplicity and is neglected because R_{sh} makes I_{sh} very small compared to the other terms. Then,

$$I = I_L - I_D[\exp(V/V_t - 1)] \tag{2.6}$$

Equation (2.6) only holds at the cell operating at one particular insolation and temperature of the cell.

The plot of the cell current I versus cell V gives the curve, as shown in Figure 2.26. The relationship between the irradiance represented by I_L and the V_{oc} is logarithmic. To obtain a series of curves at different irradiances and different temperatures the Jet Propulsion Lab model is used (see Section 2.4.7).

2.4.6 Open-circuit voltage and short-circuit current

The short-circuit current and the open-circuit voltage are the two most important parameters used to describe cell performance. The V_{oc} can be obtained from Eq. (2.6) by setting $I=0$ in the equation and solving for V_{oc}. Then

$$V_{oc} = AkT/q \times Ln[I_L/I_D + 1] \tag{2.7}$$

The I_{sc} is given by $I_{sc}=I_L$, by neglecting the small diode current and the shunt current under a no-load condition. The maximum voltage is produced at V_{oc} and the maximum current is generated at short-circuit current.

2.4.7 Jet Propulsion Lab model

The Jet Propulsion Lab (JPL) model is used to obtain expression under nonreference irradiance and temperature conditions; thus, it is used to generate $I-V$ characteristics at different temperatures and irradiances. The cell in Eq. (2.3) can be rewritten as

$$I = I_{sc}.[1 - [c_1.\exp[v/(c_2.V_{oc})] - 1] \tag{2.8}$$

$$C_2 = [V_{mp}/V_{oc} - 1/\ln 1 - I_{mp}/I_{sc}] \tag{2.9}$$

$$C_1 = \{1 - [I_{mp}/I_{sc}]\}.\exp\{-V_{mp}/C_2 V_{oc}\} \tag{2.10}$$

where,

V_{mp} = voltage at maximum power
I_{mp} = current at maximum power

The model shifts any (V_r, I_r) reference point of the reference $I-V$ curve to a new point $(V-I)$. The transformation can be carried out as follows:

The reference current, I_r, and the reference voltage, V_r, holds at a particular reference temperature, T_r, and reference irradiance, S_r.

The change in temperature (T = cell temperature) is given by Eq. (2.11):

$$\Delta T = T - T_r \tag{2.11}$$

The change in current caused by the change in temperature and irradiance is modeled as

$$\Delta I = \alpha[S/S_r]\Delta T + [S/S_r - 1]I_{sc} \tag{2.12}$$

Similarly, the change in voltage is given as:

$$\Delta V = -\beta\Delta T - R_s\Delta I \tag{2.13}$$

Therefore, the new cell voltage and current at any point in the reference $I - V$ curve is given by

$$V = V_r + \Delta V \tag{2.14}$$

$$I = I_r + \Delta I \tag{2.15}$$

The cell temperature, T, obeys the linear equation:

$$T = T_A + mS \tag{2.16}$$

where

T_A = ambient temperature of the cell
$M = 0.02 \; Cm^2/w$

The nonreference power, P, of the cell at different intensities can be written as

$$P = I \times V \tag{2.17}$$

With this model, the power at any irradiance and temperature can be obtained.

Calculation of R_s and R_{sh}

The classical solar cell I-V curve is extended on both sides and the R_s and R_{sh} are calculated as shown in Figure 2.29.

EXAMPLE 2.1

A photovoltaic module has the following parameters: I_{sc} = 2.5A, V_{oc} = 21 V, V_{mp} = 16.5 V, I_{mp} = 2.2 (given at standard conditions). The length of the module is L = 90.6 cm and the width of the module is W = 41.2 cm.

a) Calculate the fill factor, FF.

$$FF = I_{mp}.V_{mp}/I_{sc}.V_{oc} = 16.5 \times 2.2/21 \times 2.5 = 0.69$$

b) The module is connected to a variable resistive load at an irradiance of S = 900w/m^2. The voltage across the resistance was V = 17.5 V and the current

(Continued)

EXAMPLE 2.1—cont'd

flowing in the resistance was $I = 1.4A$. Calculate the module efficiency, η, where A is the cross-sectional area of the module.

$$A = L \times W = 90.6 \times 41.2 = 3732.72 \text{ cm}^2 = 0.373272 \text{ m}^2$$

$$H = P_o/P_{in} = I^*V/S \times A = 17.5 \times 1.4/0.373272 \times 900 = 8.103\%$$

c) The irradiance increased from $S = 900$ w/m² to $S = 1000$ w/m². How much would you increase or decrease the load resistance to extract maximum power from the photovoltaic module?
  - The load resistance at maximum power is $R_{mp} = V_{mp}/I_{mp} = 16.5/2.2 = 7.5$ ohms
  - The load resistance at $S = 900$ is $R = V/I = 17.5/1.4$ 12.5 ohms
  - The load resistance should be decreased by $R_{mp} - R = 12.5 - 7.5 = 5$ ohms

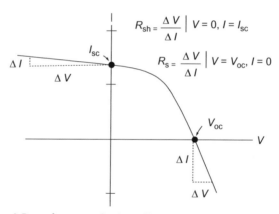

FIGURE 2.29 R_s and R_{sh} resistances of solar cell.

2.5 CONNECTIONS OF PHOTOVOLTAIC DEVICES

2.5.1 Series connection of photovoltaic devices

A photovoltaic device could be a cell, module, panel, or an array. Photovoltaic devices are connected in series to increase the voltage (Figure 2.30).

Two similar photovoltaic devices connected in series

Voltage: The voltage, V, across two similar series-connected PV devices, A and B, is equal to the sum of the individual device voltages.

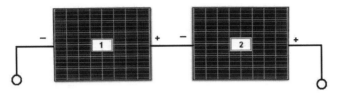

FIGURE 2.30 Two similar PV devices connected in series.

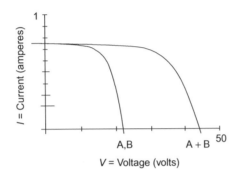

FIGURE 2.31 *I–V* curves for two similar PV devices connected in series.

$$V = V_1 + V_2 \tag{2.18}$$

$V_1 = V_A$
$V_B = V_2$

Current: The current, I, through the two similar series-connected PV devices is equal to the current through either of the individual devices. (See Figure 2.31.)

$$I = I_1 = I_2 \tag{2.19}$$

Two dissimilar PV devices 1and 2 connected in series

Voltage: $\qquad\qquad\qquad V = V_1 + V_2 \tag{2.20}$

Current: $\qquad\qquad\qquad I = I_2,\ \text{where } I_2 < I_1 \tag{2.21}$

See Figures 2.32 and 2.33 for graphic representation of the equations.

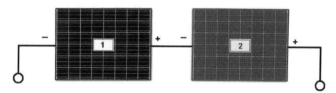

FIGURE 2.32 Two dissimilar PV devices connected in series.

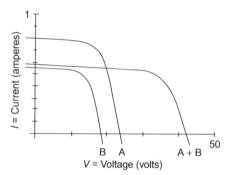

FIGURE 2.33 *I–V* **curves for two dissimilar PV devices connected in series.**

Any number of photovoltaic devices connected in series

Voltage:

$$V = V_1 + V_2 + \ldots V_N \tag{2.22}$$

Current:

For similar devices:

$$I = I_1 = I_2 = \ldots = I_N \tag{2.23}$$

For dissimilar:

$$I = I_{MIN} \tag{2.24}$$

See Figure 2.34 for graphic representation of the equations.

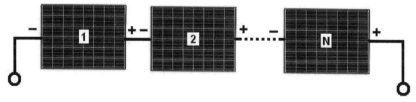

FIGURE 2.34 *I–V* **curves for two dissimilar PV devices connected in series.**

2.5.2 **Parallel connection of photovoltaic devices**

Photovoltaic devices are connected in parallel to increase the current.

Two similar photovoltaic devices connected in parallel

Voltage: The voltage across two similar parallel-connected PV devices is equal to the voltage across either of the individual devices (Figure 2.35).

$$V = V_1 = V_2 \qquad (2.25)$$

Current: The current through two similar parallel-connected PV devices is equal to the sum of the Individual device currents (Figure 2.36).

$$I = I_1 + I_2 \qquad (2.26)$$

Two dissimilar photovoltaic devices connected in parallel

Voltage:

$$V = \tfrac{1}{2}\,(V_1 + V_2) \qquad (2.27)$$

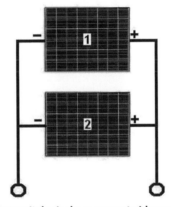

FIGURE 2.35 Two similar photovoltaic devices connected in parallel.

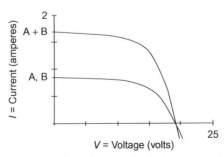

FIGURE 2.36 *I–V* curves for two similar photovoltaic devices connected in parallel.

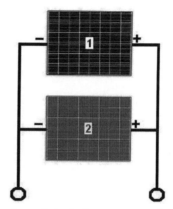

FIGURE 2.37 Two dissimilar photovoltaic devices connected in parallel.

Current:

$$I = I_1 + I_2 \tag{2.28}$$

See Figures 2.37 and 2.38 for graphic representation of the equations.

Any number of photovoltaic devices connected in parallel
Voltage:

For similar devices

$$V = V_1 = V_2 = \ldots = V_N \tag{2.29}$$

Dissimilar devices

$$V = 1/N \, (V_1 + V_2 + \ldots + V_N) \tag{2.30}$$

Current:

$$I = I_1 + I_2 + \ldots + I_N \tag{2.31}$$

See Figure 2.39 for graphic representation of the equations.

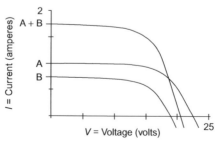

FIGURE 2.38 *I–V* curves for two dissimilar PV devices connected in parallel.

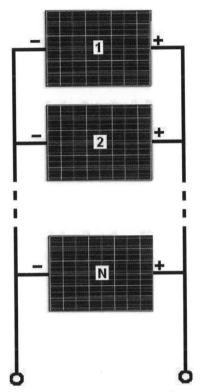

FIGURE 2.39 Any number of photovoltaic devices connected in parallel.

2.5.3 **Blocking diode**

A blocking diode is placed in series with the PV device to prevent reverse flow of current into the any number of modules (Figure 2.40). This is crucial when the load includes a battery or another source of power.

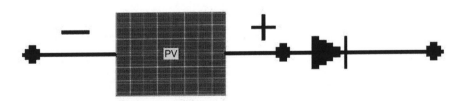

FIGURE 2.40 Blocking diode diagram.

2.5.4 Bypass diode

The bypass diode is placed in parallel with the photovoltaic module or panel (Figure 2.41). A bypass (or shunt) diode allows current to bypass the module (or group of cells) or panel in the event of an open-circuit condition or failure.

2.5.5 Orientation of collector surface

To maximize the power output of a solar device (module, panel, or array), the surface of the device should be perpendicular to the sunbeams. Sun-tracking techniques can be used to do that, but that adds to the cost. If the device is to be fixed all the time, then the tilt angle should be equal to the latitude of the location (Figure 2.42). This optimizes the energy output.

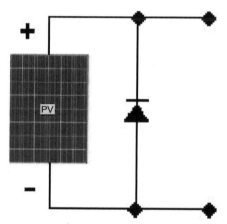

FIGURE 2.41 Bypass diode diagram.

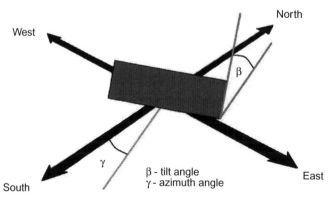

β - tilt angle
γ - azimuth angle

FIGURE 2.42 Orientation of collector surface.

EXAMPLE 2.2

Two dissimilar PV modules are subjected to the same irradiances and temperatures.

- The short-circuit current output of the first module, $I_1 = 3A$, and the open-circuit voltage output, $V_1 = 24$ V.
- The short-circuit current output of the second module is $I_2 = 2$ A and the open-circuit voltage output is $V_2 = 16$ V.
    a) The two PV modules are connected in series, as in Figure 2.35. Calculate the combined short-circuit current output, I, and the combined open-circuit voltage across the both, V:
    $I = I_2 = 2A$, the smallest of the two output currents
    $V = V_1 + V_2 = 24 + 16 = 40$ V, the sum of the two output voltages
    b) The two modules are connected in parallel, similar to Figure 2.40. Calculate the combined short-circuit current, I, and the combined open-circuit voltage, V:
    $I = I_1 + I_2 = 3 + 2 = 5A$, the sum of the two currents
    $V = (V_1 + V_2)/2 = (24 + 16)/2 = 20$ V, the average of the two voltages

EXAMPLE 2.3

Two similar photovoltaic modules are subjected to the same irradiances and temperatures. The short-circuit current output of the first module is $I_1 = 3A$, and the open-circuit voltage across the module terminals is $V_1 = 18$ V. The short-circuit current output of the second module is $I_2 = 3$ A and the open-circuit voltage across its terminals is $V_2 = 18$ V.

a) The two modules are connected in series similar to Figure 2.33. Find the combined short-circuit current, I, and the combined open-circuit voltage, V:

$$I = I_1 = I_2 = 3A$$
$$V = V_1 + V_2 = 18 + 18 = 36 \text{ V}$$

b) The two modules are connected in parallel similar to Figure 2.38. Calculate the combined short-circuit current, I, and the combined open-circuit voltage, V:

$$I = I_1 + I_2 = 6A$$
$$V = V_1 = V_2 = 18 \text{ V}$$

2.6 OPTIMIZATION OF PHOTOVOLTAIC ARRAYS

2.6.1 Introduction

The problem with using PV arrays is that variations in solar irradiance and temperature will nonlinearly affect the current–voltage output characteristics of the array. As shown in Figure 2.43, the current–voltage curve will move and deform depending on irradiance level and temperature. Consequently, the power curve for the array will shift and deform, as in Figures 2.44 and Figure 2.45. For the array to be able

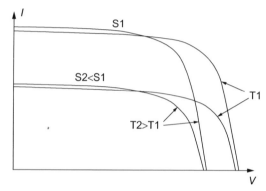

FIGURE 2.43 *I–V* curve for PV array.

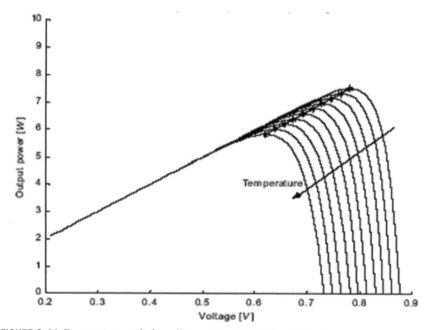

FIGURE 2.44 Temperature variation affects power curve of a solar array.

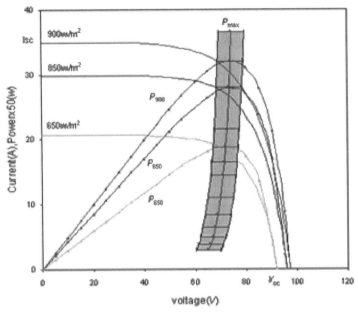

FIGURE 2.45 Maximum power points on *I–V* curves.

to output the maximum possible amount of power, either the operating voltage or current should be carefully controlled such that it operates at the maximum power point of the curve (refer to Figure 2.45).

Over the past two decades, various techniques have been conceived to optimize PV systems. Resistive matching, battery storage, and maximum power-point tracking are optimization techniques that force the array to operate at its maximum power point. Solar concentration, sun tracking, and electrical array reconfiguration are methods that attempt to increase the current output from the PV system.

2.6.2 Good matching method

Resistor load matching can really only be applied when the device powered by the array is used at a constant solar irradiance level. For a known irradiance level, only one resistor will force the array to operate at the maximum power point (see Figure 2.46). As long as the device uses this resistor for termination, the PV array will continue to supply maximum power. If the irradiance level changes, then the resistor must be changed to match the maximum power point.

2.6.3 Battery storage method

A battery can be used to maximize power coming from the PV array by adding it in parallel with the PV array. In this method, the battery will force the PV array to operate at its maximum power point. But the battery–bus voltage must be determined

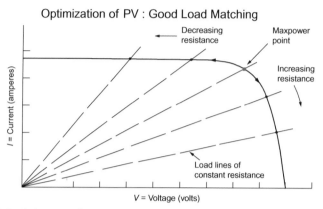

FIGURE 2.46 Resistive matching.

before using the system. The designer must know the solar irradiance level beforehand, and use a bus voltage that corresponds to the maximum power point of the array.

An approximate estimate of the maximum power point is 80% of the array's open-circuit voltage; see Figure 2.47. This method has its own flaws because of the affect of the temperature. As the temperature increases, the battery voltage increases and the PV array voltage decreases. The mismatch between the battery and the PV array is multiplied.

2.6.4 Solar concentration method

The main purpose of using solar concentrators is to increase solar radiation flux coming directly from the Sun. At low solar irradiance levels, a flat solar collector will not provide high current output. By using reflective surfaces or lenses, solar radiation can

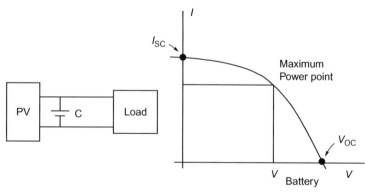

FIGURE 2.47 Battery in parallel with PV array.

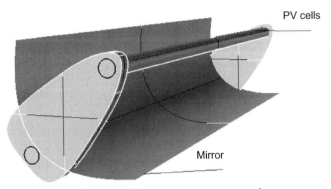

FIGURE 2.48 Parabolic trough photovoltaic concentrator: schematic.

be focused, or concentrated, before striking the PV array. Among the many solar concentrators, the popular types are the parabolic trough concentrator shown in Figures 2.48 and 2.49, the linear Fresnel lens concentrator (Figure 2.50), and the point focus Fresnel lens concentrator (Figures 2.51 and 2.52).

2.6.5 Sun-tracking method

A fixed flat-plate PV array is far from optimized when the Sun is in the West and the face of the array points toward the East. Clearly, if the plane of the array lies normal to the Sun's beams, energy absorbed by the array would be at its highest level. By using a sun tracker, power output can be improved as much as one-third. Sun trackers can be of three general categories: passive, microprocessor, and electro-optically controlled units.

FIGURE 2.49 Trough parabolic photovoltaic concentrator: actual.

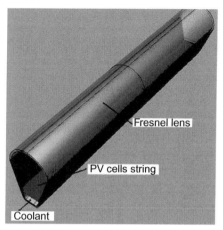

FIGURE 2.50 Linear Fresnel lens photovoltaic concentrator.

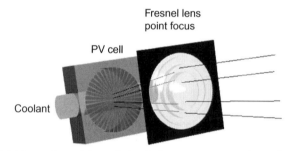

FIGURE 2.51 Point focus Fresnel lens photovoltaic concentrator: schematic.

FIGURE 2.52 Point focus Fresnel lens photovoltaic concentrator: actual.

Passive sun-tracker systems do not require any electronic controls or motors. Manufacturers (e.g., Zomeworks) build commercial trackers that work on fluid dynamic principles. These trackers consist of a fluid, such as Freon, within a frame of pipes. If the array is misaligned, one side of the frame will be heated more than the other side; as a result, the heated Freon will evaporate. Effectively, either a piston is pushed or the Freon may flow to the other side of the array, moving the array by gravity. Overall, these trackers provide moderate tracking accuracy. At times, they can be fixed in the wrong position or shift in the wind. Tracking systems that use high-torque motors do not have these problems.

Microprocessor-controlled trackers use complex mathematical formulas to determine the Sun's position. Therefore, sunlight sensors are unnecessary for this type of tracking. Stepper motors, or optical encoders, position the array such that it is aligned toward to the Sun at any time of the day. The controller calculates the Sun's position based on latitude, longitude, time, day, month, and year.

Conversely, electro-optical trackers actively search for the Sun's position dynamically using sun sensors. They are simpler and lower in cost than microprocessor-controlled trackers; in addition, they do not have to be periodically recalibrated. One system, designed by Lynch and Salameh, uses four photo resistors with cylindrical shades as a sun sensor. Its controller is designed using differential amplifiers, comparators, and output components.

This sun-tracking system is made up of eight subsystems as shown in Figure 2.53. The primary sensor is a four-sided pyramid with a solar cell mounted on each face (see Figure 2.54). It provides information on the relative position of the Sun. The two-element fixed beam sensor is designed to block as much diffuse light as possible (see Figure 2.55). It senses the presence of a sunlight beam and provides information on the absolute location. The position encoder monitors the tracker location. Usually, the track subcircuit receives information from the primary sensor and, from this data, it controls the motion of the mechanical drive system consisting of the azimuth and zenith motors.

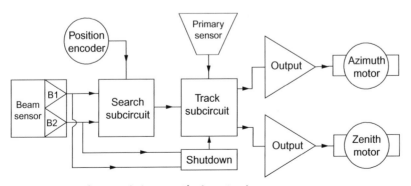

FIGURE 2.53 Block diagram of electro-optical sun tracker.

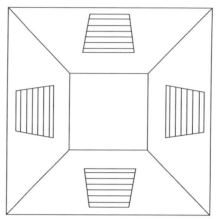

FIGURE 2.54 Pyramid sensors.

If the beam sensor does not detect a sunbeam, the shutdown subcircuit will override the primary sensor control and disable the tracker. If the beam sensor and position sensor indicate the Sun is out of range of the primary sensor, the search subcircuit will also override the primary control. It will then determine the direction of motion of the azimuth drive motor. At the output stage, two common source complementary power MOSFET half-bridge switches make up the output subcircuit. Finally, azimuth and zenith movements are controlled with DC gearbox servo motors.

In the primary pyramid sensor, the two pairs of solar cells are wired in opposing series. Each solar cell in the pair acts as a high-impedance load to the other, providing increased sensitivity. When the track mode is active, the tracker moves in a direction to where light intensity is equal on all four primary sensor cells. By adding a beam sensor to the design, light tracking in diffuse solar irradiance conditions is improved. As shown in Figure 2.55, phototransistors are located at the center of each half of the divided dome.

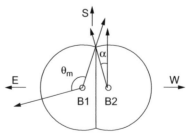

FIGURE 2.55 Beam sensors.

2.6.6 **Electrical array reconfiguration method**

In a PV powered DC motor system, low solar irradiance levels can unfavorably turn off the motor. Since the torque of the motor is proportional to the armature current $(T = K* \phi *I_a)$, low PV output current will reduce the torque below the stall torque necessary to keep the motor running. A two-stage electrical array reconfiguration controller (EARC) is designed to keep the motor running by switching between a parallel and a series PV array configuration, based on insolation level.

In the early morning, late evening, and on cloudy days, solar irradiance on the PV array is low. At such times, the maximum current available for starting the motor is roughly the PV array's short-circuit current I_{sc}. To increase the available current, two PV array units, U1 and U2, can be reconfigured so that they are wired in parallel (see Figure 2.56). During periods of high solar irradiance, the system will be reconfigured so that U1 and U2 are wired in series (see Figure 2.57).

The switching circuitry of the two-stage EARC is designed with three N-channel enhancement mode metal oxide silicon field effect transistors (MOSFETs). In high irradiance, MOSFET G1 (see Figure 2.58) will be active so that PV array units are in series. In this configuration, current supplied to the motor is I and the voltage is V. When the two PV units are illuminated with low irradiance, MOSFETs G2 and G3 are active so that the PV array units are in parallel. So, the current supplied to the motor is 2I (twice the current in series mode), and the voltage is V/2 (half the voltage in series mode). Since the permanent magnet motor has a wide operating voltage, this reduction in voltage (i.e., parallel mode) will not shut off the motor. With the addition

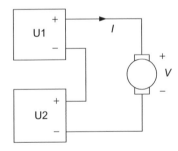

FIGURE 2.56 Serial mode.

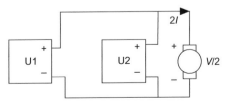

FIGURE 2.57 Parallel mode.

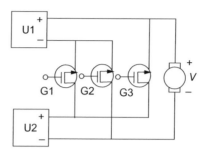

FIGURE 2.58 Electrical array reconfiguration for a switching circuit.

of a reference solar cell, a threshold irradiance level can be set to switch between parallel and serial modes.

A reference solar cell, mounted in the same plane and direction as the PV array, will produce a current proportional to irradiance level. In Figure 2.59, this reference current is transformed into a voltage through a 47-ohm load resistor and compared with a set threshold voltage. Two comparators are necessary to turn on either G1 (series mode) or G2 and G3 (parallel mode). When the reference voltage is greater than the set point, the higher comparator will trigger G1; effectively, PV arrays U1 and U2 will be wired in series. Conversely, when the reference voltage is less than the

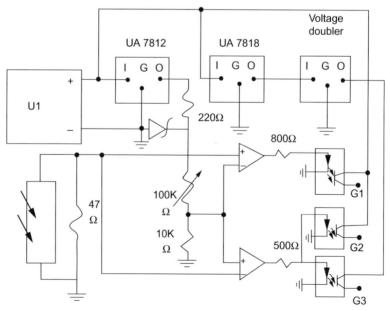

FIGURE 2.59 Electrical array reconfiguration for a controller circuit.

set point, the lower comparator in the figure will trigger G2 and G3; therefore, PV arrays U1 and U2 will be wired in parallel.

As seen in Figure 2.59, the reference voltage for setting the insolation set point is obtained with a UA7812 (12-volt regulator), a 220-ohm resistor, and a 3.6 V Zener diode. When the regulator falls out of regulation (i.e., when the PV array output voltage is low), the Zener diode will keep the reference node at 3.6 V. The 100 kΩ potentiometer is used to adjust the set point. To simplify the triggering circuit, optoisolators are used to provide the MOSFET gate voltage. Since MOSFET G3 will have a source and drain voltage as high as 24 V, a voltage doubler is necessary to supply a gate voltage high enough for turning on G3. Notice how all regulators in the EARC are powered by the PV array. Therefore, an external voltage source, such as a battery, is not required.

2.6.7 Maximum power-point tracker method

Since modern PV arrays still have relatively low conversion efficiency, the overall PV system cost can be reduced using high-efficiency power conditioners, DC/DC switch-mode converters, for load interface. When building a maximum power-point tracker (MPPT), a DC/DC converter will typically be controlled with analog circuitry or a microcontroller (Figure 2.60).

To extract maximum power from the PV array, an MPPT is used to control the fluctuating operating power point of the array. As stated earlier, variations in insolation and temperature will change the operating power point to a value less than maximum. Rather than optimizing the current or voltage at a particular irradiance or temperature, an MPPT continuously tracks the maximum power point of the PV array. Either the operating voltage or the operating current should be carefully controlled so that the maximum power is sourced by the array.

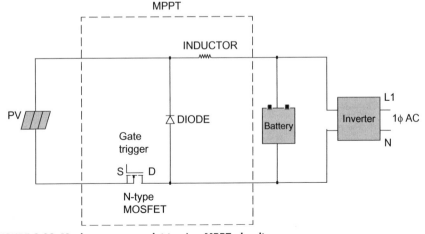

FIGURE 2.60 Maximum power-point tracker MPPT circuit.

Most current MPPT designs consist of two basic components: a switch-mode converter and control circuitry. The switch-mode converter is a DC/DC converter that allows energy at one potential to be drawn, stored as magnetic energy in an inductor, and then released at a different potential. By configuring the DC/DC converter in various topologies, a high-to-low voltage (buck or step-down) or low-to-high (boost or step-up) voltage converter can be constructed.

MPPT control algorithms

Many algorithms have been devised to control the MPPT. Three of the most commonly used techniques are the perturbation and observation method, incremental conductance algorithm, and power slope method.

The perturbation and observation method operates by periodically perturbing the array terminal voltage and comparing the PV output power with that of the previous perturbation cycle. If the voltage perturbation results in an increase in PV system output power, the subsequent perturbation is made in the same direction. Conversely, if the voltage perturbation results in a decrease in PV output power, the succeeding perturbation is made in the opposite direction. Clearly, if the voltage perturbation does not change the PV output power, no change is made in the following cycle.

For the incremental conductance algorithm, two calculated values are necessary to move toward the maximum power point: the source conductance ($G = I/V$) and the incremental conductance ($\Delta G = dI/dV$). The goal of this algorithm is to search for the voltage operating point at which the conductance is equal to the incremental conductance.

The output power from the source can be expressed as

$$P = V \times I \tag{2.32}$$

Applying the chain rule for the derivative of products yields

$$dP/dV = d(V \times I)/dV \tag{2.33}$$

$$= I \times dV/dV + V \times dI/dV \tag{2.34}$$

$$= I + V \times dI/dV \tag{2.35}$$

$$\therefore (1/V) \times dP/dV = I/V + dI/dV \tag{2.36}$$

Source conductance is defined as

$$G = I/V \tag{2.37}$$

Then, the source incremental conductance is defined as

$$\Delta G = dI/dV \tag{2.38}$$

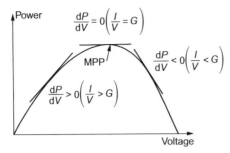

FIGURE 2.61 Incremental conductance algorithm.

In general, the output voltage from the PV array is positive; therefore, the value 1/V will be positive in Eq. (2.36). If the conductance is larger than the incremental conductance, Eq. (2.36) states that the operating voltage is below the voltage at the maximum power point, and vice versa. These ideas are summarized by Eqs. (2.39), (2.40), and (2.41) and are shown graphically in Figure 2.61.

$$dP/dV > 0, \text{ if } G > \Delta G \tag{2.39}$$

$$dP/dV = 0, \text{ if } G = \Delta G \tag{2.40}$$

$$dP/dV < 0, \text{ if } G < \Delta G \tag{2.41}$$

The power slope method uses the power slope dP/dV directly to find the maximum power point. Both the PV array output current and the voltage are sensed at consecutive time intervals so that the power slope can be calculated. The power slope is calculated as follows:

$$dP/dV = [P(\text{current}) - P(\text{previous})/V(\text{current}) - V(\text{previous})] \tag{2.42}$$

If the direction of the voltage is continuously increasing or decreasing, the power slope can be used to determine when to reverse direction. For example, if the output voltage of the PV array is increasing ($dV > 0$) and dP is calculated to be positive, then the power slope is positive. On the other hand, if dP is calculated to be negative, then the power slope is negative and the direction of voltage would need to be reversed so that the array output voltage begins continually decreasing. This "hill-climbing" algorithm, summarized in Figure 2.62, can be easily applied by controlling the duty cycle of a DC/DC converter.

Step-up MPPT controller

A block diagram of the PWM controller is shown in Figure 2.63. The array voltage and current are multiplied to obtain a value proportional to instantaneous power.

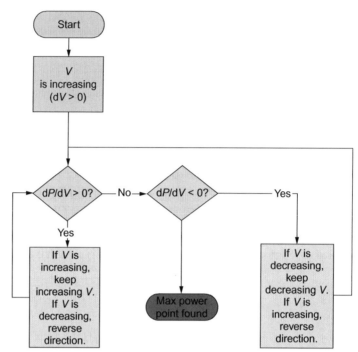

FIGURE 2.62 Flow charts for power slope algorithm.

This is fed into an RC circuit with a slow time constant and another with a fast time constant. Since an RC circuit introduces a time (phase) delay, the fast RC circuit output will be the "current" power output, while the slow RC circuit output will be the "previous" power output. In Figure 2.63, point I represents the "array current" power, while the point P represents the "previous" power. The time delay of the

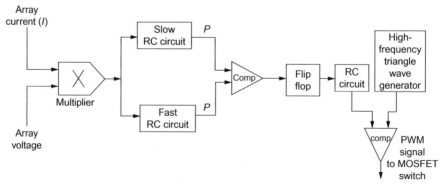

FIGURE 2.63 PWM controllers.

two points will be proportional to the difference in time constants of the two signals. If the delay is small, the error in tracking the maximum output power will be extremely low.

The output of the comparator is used to clock a flip-flop. The flip-flop is set up so that it can only charge or discharge the RC circuit at its output. As a result, the duty cycle created by comparing a high-frequency triangle wave with the voltage of the RC circuit will be either increasing or decreasing, and it can be forced to change direction only. This ensures that the duty cycle is always moving toward the maximum power point.

When the PWM controller is activated, the duty cycle will start to increase, thus increasing the power output (see Figure 2.64). Point *I* will move up the power curve some small time interval faster than point *P*. As point *I* passes the maximum power point, it will at some time be smaller than the power at point *P*. At this moment, the comparator output will switch to high and clock the flip-flop on the positive edge. As a result, the direction of the duty cycle will be reversed. So, point *I* will "climb" the power curve in the reverse direction until it once again passes point *P*. This oscillating movement continues infinitely and eventually oscillates in the small vicinity of the maximum power point.

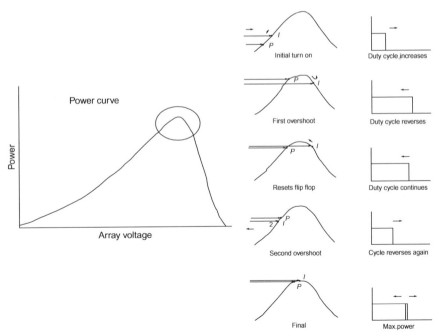

FIGURE 2.64 Duty cycle starts to increase, thus increasing the power output.

2.7 PHOTOVOLTAIC SYSTEMS

Early applications of solar cells were in space. Beginning in 1958, almost every long duration space missions undertaken by the United States and the former Soviet Union was powered by PV. The terrestrial use of photovoltaic systems has been growing steadily. During the brief four-decade history of PV, costs have come down immensely and the efficiencies have nearly tripled.

2.7.1 Calssification of photovoltaic systems

PV systems are different from conventional fossil fuel burning energy systems. While the required input of the conventional system (fuel) depends on the output, the input in a PV system depends on the insolation. PV outputs can vary as a result of external factors such as moving clouds. PV systems can be very small (less than 5 W), small (5 W–1 kW), Kilowatt size (1 kW to a few 10s of kW), and intermediate size (10s of kW–100 kW), or a large-scale system (1 mW or larger) which is connected to a utility grid for commercial or utility power generation.

There are two types of PV systems: (1) grid-connected (grid-interactive) and (2) stand-alone. There are two possible versions of a stand-alone system, depending on the load: power needs, primarily lighting, communication, entertainment, and resistive loads.

- Stand-alone PV system with DC loads; battery storage usually is required.
- Stand-alone PV system with AC loads; battery storage is usually required.
  (All have typical loads including induction motors.)

Battery storage is essential to the success of stand-alone PV systems' design. During the daylight, PV arrays charge the batteries so that they may supply energy at night and on cloudy days.

PV Grid-connected system

The PV grid-interactive or as it is called sometimes, the utility-interactive PV system, is used in residential systems, for industrial systems, and at central power stations (Figure 2.65). However, to connect a PV system to the utility, the inverter has to meet certain standards.

Stand-alone PV system with a DC load

The voltage regulator disconnects the PV when the battery voltage reaches its maximum permissible value. It also disconnects the load if the battery voltage reaches its minimum allowed value to avoid destroying the battery. A DC/DC converter is used to step up or down the PV voltage to the level of the load voltage (Figure 2.66). Batteries are used to add reliability to the system. The load can be powered even if there is no PV power. For loads (e.g., irrigation pumps) for which the time of utilization is not important, there is no need for battery storage.

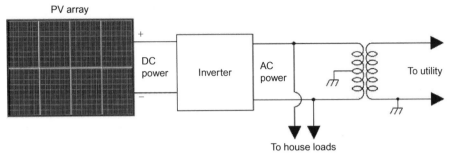

FIGURE 2.65 PV grid-connected system.

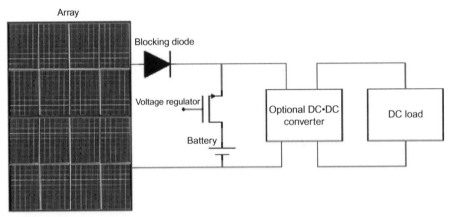

FIGURE 2.66 PV system with a DC load.

Stand-alone PV system with AC load

This PV system is supplied with an inverter to convert the DC output of the PV system to AC. An example of that is supplying a house with electricity in a remote area where there are no AC powerlines. This kind of system requires battery storage (Figure 2.67). On the other hand, powering a water pump that is driven by an AC motor rated more than 15 HP in an agriculture area does not require battery storage.

2.7.2 Components of a PV system

A photovoltaic system usually consists of a photovoltaic array, a battery bank (if required), and related balance-of-system (BOS) components designed to safely meet the power requirements of a given load.

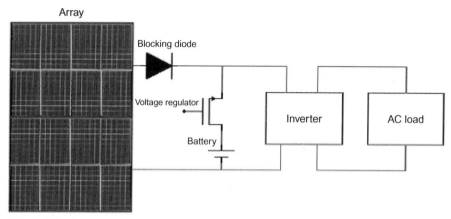

FIGURE 2.67 Stand-alone PV system with AC load.

Balance-of-system components

Balance-of-system design involves the design and selection of additional components such as:

> *The voltage regulator*, or charge controller: This is an important device when using batteries. The regulator controls the current and voltage input to the batteries and protects them from damage.
>
> *The power conditioner* (i.e., inverter or a DC/DC converter): This provides the interface between the PV system and the AC loads. It should be electrically matched to the array output and compatible with all loads. It also should meet IEEE standards.
>
> *Alarms and cut-offs* (disconnects): They are often used in PV systems to warn of low voltage across the battery bank and to disconnect a load to protect it from damage. When using batteries, it is absolutely essential that a fused disconnect be provided at the battery bank for protection.
>
> *Diodes*: These are electrical devices that are used to both block current flow from the battery bank to the PV array (at night) to protect the modules. This kind of diode is called a blocking diode. Bypass diodes, placed across the modules, are also used to allow current to bypass inoperative modules (in a series string) of a large array.
>
> *Varistors* (for surge protection), disconnects, and fuses (for over current protection) should be used throughout the system.
>
> *Metering*: A minimum of 2 kWh meters will most likely be required in a utility-interactive PV residence: one to measure utility energy flow into the residence and the other to measure energy flow back to the utility distribution network.
>
> A low-cost improved metering strategy would add one (or two) additional kWh meter(s): one to measure the AC energy output of the PV system

(and possibly a second to measure the heating and cooling energy used). This simple metering scheme would provide useful information on PV system performance (and on the major house loads). The total house load could easily be computed by adding the PV output to the utility energy used and then subtracting the excess energy fed back to the utility. The PV system is also equipped with ampermeters to measure the currents and volt meters to measure voltages.

Mounting structure: The frame to which the array is mounted could be made of wood or steel. The frame could be fixed at an angle of the latitude of the location, or track the Sun with single-axis sun tracking or two-axis tracking. Sun trackers cost more, but the efficiency of the PV system is improved. The output could be increased by more than 20%.

Rack and ground mounting: The following conditions must be taken into account when mounting PV arrays:

- Rack-mounted arrays are above and tilted at a nonzero angle to the mounting surface (roof or ground). The optimum tilt angle is usually equal to the latitude of the location where the PV system is installed
- Optimum orientation (i.e., azimuth and tilt) should be used for rack mounts.
- Rack mounts usually are subjected to higher structural loads, incur higher costs for mounting hardware, and are less attractive than stand-off mounts. However, for the same array area, the total energy output is often somewhat higher because of optimized orientation and lower average operating temperatures.
- Ground mounts should be effectively secured to the ground to resist uplifting caused by wind loads. The rack mount could be wooden or metallic structure depending on what is available
- To reduce corrosion, avoid to the extent possible the use of dissimilar metals in the framing, mounting, and grounding materials. Rack mounts (Figure 2.68) are recommended for flat roofs and ground mounts.

Characteristics of grid-interactive inverter

Because photovoltaic modules and arrays generate DC electrical energy, and most residential loads work on AC energy, PV systems need a device to convert DC electrical energy to AC electrical energy. Such an electronic device is called an *inverter*. In a grid-interactive photovoltaic system, the inverter also functions as the control—or brain—of the system, allowing power or energy to flow to and from the grid.

A properly designed inverter should also ensure the safety of the system, maximize AC energy generation consistent with the array operating at the maximum power point, ensure high conversion efficiency, and deliver high power quality. The following are the DC inverter characteristics:

DC nominal voltage: This is the design input voltage of the inverter. Generally, inverter efficiency is highest at the nominal voltage. The array voltage at

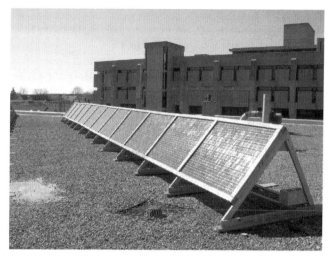

FIGURE 2.68 Rack mount.

maximum power (at SOC) should be compatible with the inverter input DC nominal voltage.

DC voltage thresholds: Inverters should be capable of operating over a range of DC input voltages to allow for operation under various expected meteorological conditions and with different array types and configurations. Lower and upper ends of the voltage range are known as the *voltage thresholds.*

DC power rating: This is the full input power rating of the inverter. The DC power rating of the inverter should exceed the array's maximum power at full-sun conditions (1 kW/m^2) and normal operating cell temperature (NOCT). In addition, the inverter should be capable of handling occasional higher insolation conditions (1.3–1.4 kW/m^2) on the array, either by limiting input power to the inverter by deviating from array maximum power tracking (MPT), or by oversizing the inverter with respect to the array power rating.

DC power threshold: The maximum DC input power to the inverter, up to which no AC power is generated, is called the DC power threshold, which results in a constant power loss in the inverter's circuitry under operation. Consequently, it should be kept as low as possible, preferably within 2% of the DC power rating.

Maximum power tracking: To obtain maximum energy output from the array under field operating conditions requires that the inverter maintain operation at, or track, the maximum power point (on array *I–V* characteristics). The voltage at the maximum power point varies with meteorological conditions, being low at low insolation and high cell temperature and high at high insolation and low cell temperature. This variation in maximum power point voltage may be ±5 to ±12% of the array SOC voltage depending on local meteorological

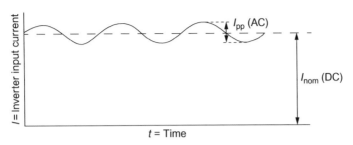

FIGURE 2.69 Ripple current.

conditions. A MPT range of 85 to 115% of the nominal input voltage is generally satisfactory. A maximum power-tracking voltage range narrower than that may cause some loss of available power/energy.

Ripple factor: This refers to the injection of alternating current from the output side of the inverter to the input, or direct current side of the inverter. This injection of AC current may be caused by electromagnetic pickup, or by conduction through the inverter chassis, grounding connections or inverter components. The injection of AC current causes the array output/inverter input current to be modulated. The ripple factor is defined as the peak–to–peak value of the AC component of the current at the inverter input divided by the nominal DC input (see Figure 2.69). A ripple of up to 5% is generally considered acceptable for power conditioners.

$$\text{Ripple } \% = \frac{I_{pp}(\text{AC})}{I_{nom}(\text{DC})} \times 100 \qquad (2.43)$$

The following are the AC inverter characteristics:

AC nominal voltage: This is the design output voltage of the inverter. For grid-interactive inverters, the AC nominal voltage is the utility voltage, which is either 240 V (phase-to-phase) or 120 V (phase-to-ground) at the distribution transformer, which is secondary for single-phase inverters.

AC power rating: This is the full output power rating of the inverter and equals the product of the DC power rating and the inverter efficiency at full power. The AC power rating is based on the inverter's AC power capability at DC nominal input voltage rating and at AC under-voltage threshold.

Utility voltage thresholds: To protect the utility distribution system, connected appliances and loads—and itself, the inverter—should operate only if the utility voltage at the inverter output interconnection point is within an acceptable range. The lower and upper values of this acceptable range are known as the utility or AC (output) voltage thresholds. The recommended under-voltage and over-voltage thresholds are 86.7% and 105.8% of the AC nominal output voltage, respectively.

Operational frequency range: The output frequency of the grid-interactive inverter must be synchronous with the utility. Also, to protect the utility distribution system, connected loads, and appliances, the inverter should operate only in a narrow range of utility frequencies; the recommended range for inverter operation is from 59.5 to 60.5 Hz (by Sandia National Laboratories). This narrow range of operational frequencies has been recommended to alleviate some of the concerns of utilities.

Total harmonic distortion (THD): The utility service voltage is nominally sinusoidal at 60 Hz. The inverter output voltage and current waveforms may deviate, however, from a purely sinusoidal 60 Hz signal. Any distorted waveform can be shown analytically to consist of a 60 Hz (fundamental) frequency and higher harmonic frequencies (multiples of 60 Hz), as shown in Figure 2.70. The square root of the sum of the squares of the amplitudes of the individual harmonics divided by the amplitude of the fundamental frequency is called THD of the waveform (current or voltage):

$$\text{THD} = \frac{\sqrt{\sum_{n=2}^{\infty} A_n^2}}{A_1} \tag{2.44}$$

where A_n is the amplitude of nth harmonic.

The current harmonics injected onto the utility grid also produce voltage harmonics with magnitudes that are dependent on the utility system impedance at the harmonic frequencies. Because the utility system impedance is usually very low, the voltage THD is much lower than the current THD.

Harmonic distortion can be detrimental to connected loads and utility equipment because the higher-frequency components generate heat in the

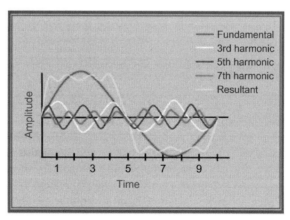

FIGURE 2.70 Harmonics in distortion waveforms.

transformer, induction motors, and other loads. This and the possibility of resonance from the combination of utility reactive components (capacitors and inductors) at harmonic frequencies have led to recommended limits on inverter output harmonic distortion. For a good quality grid-interactive inverter, the current THD should be less than or equal to 5% and the voltage THD should be less than or equal to 2%. Single-frequency harmonic components should be less than or equal to 3% for current distortion and less than or equal to 1% for voltage distortion.

Power factor: This is defined as power supplied divided by the product of current and voltage output. If the voltage and current output are in phase in Figure 2.70, $\theta = 0$, then the power factor is unity and all the volt-amperes are used to do work. If the power factor is less than unity, then the volt-amperes output of the system is greater than the useful power measured in watts. With a lower power factor, not only do the transmission losses increase but also the utility system has to be designed to handle more volt-amperes, thus requiring oversized distribution system components. Also, a lower power factor may affect voltage regulation. The recommended range of power factors for grid-interactive inverters is from 0.95 lagging (inductive) to 0.95 leading (capacitive). (See Figure 2.71.)

- AC circuits: V wave is not always in phase with I wave
- Power factor $=$ cosine θ, power $= VI$ cosine θ

Utility isolation: Electrical isolation between the inverter AC output and the utility service interconnection point is necessary to prevent DC current from the array from reaching the distribution transformer or other connected loads. For this purpose, an isolation transformer is recommended in the inverter at the output stage.

Electromagnetic interference (EMI): High-frequency switching (60 Hz to 100 kHz) in the inverter circuitry has the potential of producing electromagnetic interference. EMI may adversely affect communication equipment, such as telephones, radios and televisions (in the radiated interference mode), or any equipment fed by the same utility system (in the conducted interference mode).

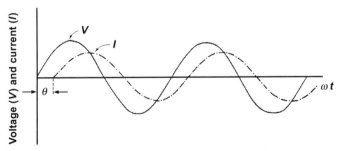

FIGURE 2.71 AC circuit current and voltage waveforms.

AC standby power: The AC standby power is the utility AC power consumed by the inverter circuitry when no AC power is being generated at the inverter output. It is recommended that the inverter have an automatic "night switch" so that utility power consumption is minimized when the array output is lower than the inverter DC power threshold, which occurs at low insolation and at night.

Audio noise: The switching circuitry and magnetic components in the inverter generate acoustical noise. For this reason, the inverter should be located in a less sensitive area of the residence. The use of sound insulation may be warranted with some inverters; however, care should be exercised not to inhibit ventilation. A maximum sound level of 0.75 dbA (reference 0.0002 microbar) at a distance of two meters from the surface of the inverter cabinet has been recommended. The equivalent of a General Radio Type 1551B sound level meter should be used for measurements.

2.7.3 Design of a stand-alone PV system with battery storage

The design of stand-alone systems is readily accomplished by systematically considering the load to be powered, the type and size of batteries required, photovoltaic array size, additional electrical and structural components, and cost.

Load evaluation

Evaluation of the load (i.e., the energy and power requirements of the equipment to be served by the PV system) is the first step in the design process. Equipment to be powered by the system should be as energy efficient as possible. In lighting applications, for example, fluorescent or sodium vapor lights should be selected over less efficient incandescent. Using DC equipment, thereby eliminating the inverter, results in more efficient utilization of energy.

The end result of the load evaluation should be a load profile over 24 hours for a typical day of each month. This will be required in order to size both the battery bank and the PV array. In summary, load evaluation is accomplished as follows:

1. Determine equipment power and energy requirements (watts and watt-hours)
2. Specify equipment electrical characteristics (voltages, AC or DC)
3. Evaluate and/or improve equipment efficiencies
4. Determine load profile

Battery storage sizing

The battery storage system is sized independent of the photovoltaic array. The equipment load is converted from watt-hours to amp-hours at the system DC voltage in order to size the battery bank. The battery system is sized for the worst case and should be assumed to carry the entire load for a fixed number of days. The days of storage to use is a subjective judgment arrived at by considering "actual conditions" and how much they may vary from "average conditions" at the location. Generally, this is a site-specific variable and typical storage periods range from 3 to 10 days.

Battery capacity, in units of amp-hours, is temperature-dependent and manufacturers' ratings should be corrected for temperature in order to serve the load as intended. Once the required number of amp-hours has been determined, batteries or cells can be selected and the battery bank designed using manufacturers' information and desired depth of discharge. The total number of batteries needed is the product of the number of series batteries (to build up the voltage) and the number of parallel batteries (to build up the capacity). In summary, battery storage sizing is accomplished as follows:

1. Determine load amp-hours requirement
2. Correct amp-hours for temperature
3. Determine the number of batteries (cells) required

Photovoltaic array sizing

The size of the PV array is determined by considering the available solar insolation, the tilt of the array, and the characteristics of the photovoltaic modules being considered. The available insolation striking a photovoltaic array varies throughout the year and is a function of the tilt of the array. If the load is constant, the designer must consider the time of year with the minimum amount of sunlight (typically December or January). Knowing the insolation available (at tilt) and the power output required, the array can be sized using module specifications supplied by manufacturers.

Although it is energy (watt-hours) that is consumed by the loads, PV modules are current-producing devices. Voltages are established by the load or battery bank. Consequently, in systems where the voltage is permitted to vary, the designer should consider current (rather than voltage) in sizing the array. Both "peak" power and current (at peak power) are available from manufacturers' specifications.

Using module peak current and daily insolation (in peak sun hours), the average amp-hours of current delivered by a PV module for one day of the "worst" month can be determined. The amp-hours required by the load determined previously are adjusted (upward) because batteries are less than 100% efficient. (Usually 80% battery efficiency is assumed.) Then, knowing the requirements of the load and the capability of a single module at the location during the worst month, the array can be sized. In summary, photovoltaic array sizing is accomplished as follows:

1. Determine available sunlight—number of peak sun hours
2. Select array, tilt, and candidate modules
3. Size array (compute daily amp-hours output)
4. Evaluate array–battery combination

STEPS FOR SIZING A REMOTE, STAND-ALONE PV SYSTEM WITH BATTERY STORAGE

Step 1: Calculation of load or the load energy requirement

A1 Find the inverter efficiency (decimal) that will be used in the stand-alone PV system

A2 Decide the battery bus voltage, volts. Choose the electrical components to be installed to form the load.

A3 Find the rated wattage of each component.

A4 Find the adjusted wattage of each component; that is if the operating temperature is not the same as the design one.

A5 Determine the number of hours per day each component will operate.

A6 Find the energy demand per day for each component.

A7 Add all the energy demand per day of all the components—the load.

A8 Finally, find the amp-hours demand per day of all components.

For a residential house, for example, the estimated energy requirement in kWh per month for each component is given in Table 2.1. This table can be used to calculate the energy requirements per month and, if you divide it by 30, you get the energy requirement per day—A7 and A8.

Step 2: Calculation of battery storage capacity

B1 Days of storage desired and/or required. This number depends on the location of the PV system; it is higher in the North and less in the sunny places. Usually, this is the number of consecutive days expected without sun.

B2 Desired depth-of-discharge (DOD) limit of batteries to be used (decimal). This number depends on battery type. It is 80% for lead acid batteries and 100% for NiCd batteries.

B3 Required usable capacity (A8 × BI) in amp-hours.

B4 Amp-hours capacity of selected battery that is going to be installed in the PV system.

B5 Useful battery capacity (B4 × B2) in amp-hours.

B6 Number of total batteries needed (B3 ÷ B5).

Step 3: Determine the PV array sizing

CI Total energy demand per day (A7) in watt-hours.

C2 Battery round-trip efficiency (0.80–0.85) for lead acid batteries.

C3 Required array output per day (C1 ÷ C2) in watt-hours to satisfy the load energy demand.

C4 Selected PV module open-circuit voltage × 0.8 in volts. This is roughly the operating voltage of the PV module.

C5 Selected PV module power output at 1000 watts/m^2. This number is given by the manufacturer of the selected PV module

C6 Peak sun hours at optimum tilt angle, which is usually the latitude angle. This number depends on the location of the site; each location on Earth is assigned a certain number, depending on how sunny the place is.

C7 Energy output per module per day (C5 × C6) in watt-hours.

C8 Operating temperature derating factor (DF) for hot climates and critical applications (0.80); for moderate climates and noncritical applications (0.95)—C8 = (DF × C7) in watt-hours.

C9 Number of modules required to meet the load energy requirements (C3 ÷ C8).

Table 2.1 Residential Electricity Requirements

Product	Est. kWh used monthly	Product	Est. kWh used monthly
Food preparation		**Comfort conditioning**	
Broiler/rotisserie	7	Central system	
Coffee maker	9	2 ton	1450
Deep fat fryer	7	3 ton	2100
Dishwasher (incl. hot water)	90	4 ton	2750
		Dehumidifier	144
Frying pan	8	Electric blanket	12
Microwave oven	10	Fans	
Range with Oven	42	Whole house	30
Roaster	5	Ceiling	12
Slow cooker	12	Circulating	4
Toaster	3	**Water heating supply**	
Trash compactor	4		
Blender/can opener	< 1 each	Domestic supply pump	27
Mixer/waste disposal	< 1 each	Pool pump (3/4 hp)	375
Food preservation		Sprinkler system (1.5 hp)	28
		Water heater	
Refrigerator		Average use	212
Manual 12 cu. ft	78	Typical use, 2 person	195
Refrigerator-freezer		Typical use, 4 persons	310
Manual 12.14 cu. ft	125	**Health and beauty**	
Frost-free 14-17 cu. ft	170		
Frost-free 17-20 cu. ft	205	Hair dryer	2
Freezer		Hair roller/heating pad	1
Manual 14.5-17.5 cu. ft	135	Heat lamp/sun lamp	1
Frost-free 14.5-17.5 cu. ft	138	Curling iron/shaver	1
		Toothbrush	1
Laundry services		**Home entertainment**	
Dryer	78	Radio	7
Hand iron	5	Radio/music player	9
Clothes dryer	10	Television	
Lighting		B&W, tube type	18
		B&W, solid state	8
4-5 rooms	50	Color, tube type	44
6-8 rooms	60	Color, solid state	27
Outdoor spotlight, all night	45		
Household necessities			
Clock	1.5		
Sewing machine	1		
Vacuum cleaner	4		

EXAMPLE 2.4

A house has appliances and daily energy consumption, as shown in the following table.

Appliance	Daily consumption in/kWh
Coffee maker	0.3
Microwave oven	0.333
Refrigerator	4.167
Washing machine	0.33
Vacuum cleaner	0.133
Television	1.467
Blender or mixer	2
Slow cooker	8
Dishwasher	3
Toaster	0.1
Dryer	2.6
Lighting	1.667
Fan	0.4
Hair dryer	1.0
Hand Iron	5
Total	37 kWh, the house daily energy demand

a) Calculate the house's daily energy demand.
b) If the battery bus voltage is 12 volts, then calculate the house amp-hours demand—$37,000 \div 12 = 3083.3$ Ah.

EXAMPLE 2.5

The average energy demand per day for a stand-alone house is $E = 24$ kWh.

a) Calculate the number of batteries needed for 6 days of energy storage. The available batteries' specification is $V = 12$ V, capacity is $C = 450$ AH, battery efficiency is 80%, DOD = 100%—house energy demand for 6 days, E_H:

$$E_H = 6 \times 24 = 144 \text{ kWh}$$

A battery energy capacity

$$E = 450 \times 12 = 5.4 \text{ kWh}$$

Total number of batteries needed

$$144 \div 5.4 = 26.66$$

We choose the closest number, rounded to next higher integer—27 batteries

EXAMPLE 2.5—cont'd

b) If the number of peak sun hours at the location of the house is 5, calculate the number of the 100 Wp PV modules needed to meet the energy demand of the house using the previously mentioned batteries; neglect the inverter losses. House daily energy to be generated by the PV system taking into account batteries' losses:

$$E = 24 \div 0.8 = 30 \, \text{kWh}$$

The daily energy output of a single panel is:

$$5 \times 100 = 500 \, \text{kWh}$$

The number of panels $= 30,000 \div 500 = 60$ panels.

2.8 APPLICATIONS OF PHOTOVOLTAICS

The light from the Sun, in conjunction with special semiconductive material, can be used to create energy, which in turn can power electrical devices without any recurring costs or harm to the environment. As technology improves, electrical components are being made to consume less and less energy. Meanwhile, photovoltaic devices, commonly called solar cells, are being created with better fill factors and greater efficiency. This combination, along with better energy storage and lower manufacturing costs, has made solar power more practical with each passing year. Figure 2.72 shows where in the world there is solar power.

2.8.1 Solar evolution

Although PV technology was discovered much earlier, the solar revolution began back in the late 1970s or early 1980s. This was when people became aware of using the Sun for energy. It was not uncommon to see huge solar panels protruding from a neighbor's rooftop along with plumbing that circulated water. These were truly ugly and plagued with problems; they were wind catchers, the housings often leaked, and required a cooling pump. A few years later the boom was over, and these panels began to disappear from people's houses. Seldom do you see panels from the early 1980s. Although largely used for heating water, the vision of those panels is engrained in people's minds when they think of "solar panels."

The next trend was the solar calculator, which was inexpensive and effective. There must be one in every household. The solar calculator opened people's eyes to accepting solar power for small devices. Figure 2.73 shows an older style calculator that is still available today. Figure 2.74 shows a more advanced use of solar energy, where a PV system is used to power the international space station, ISS.

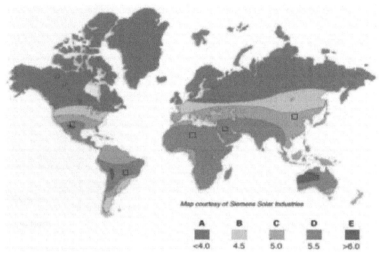

A	B	C	D	E
<4.0	4.5	5.0	5.5	>6.0

FIGURE 2.72 World map relative to solar panel use.

FIGURE 2.73 Basic solar calculator.

Products for which batteries are not required

PV technology coincides well with low power consumption devices that have a sufficiently large surface area that allows for a form-fit solar panel. Although there are limitless applications for practical devices, these are often seen on gifts, gizmos, and gadgets (Figure 2.75). The solar calculator is a perfect example of this. They have become so common because they are often used as a gimmick to advertise a company. A small watch battery could power a basic calculator for years, so there is

FIGURE 2.74 The International Space Station, ISS.

FIGURE 2.75 Toy helicopter with rotating blades.

not much of a market for selling them in stores. Companies (e.g., banks) will often buy these items in bulk, print their logo on them, and give to customers.

Some other examples include toys, such as planes, windmills, and radiometers that spin faster with increasing sunlight levels (Figure 2.76). Also, sun chimes, instead of wind chimes, have a small motor that moves the chimes when there is no wind to provide a soothing sound on a porch or in a garden (Figure 2.77). Each of these products uses only a solar panel and requires no energy storage to operate. Because no batteries needed are, these items can be made and sold quite inexpensively.

Products for which batteries are required

Some other popular, yet inexpensive, solar products include radios, flashlights, and mosquito guards (Figure 2.78). Each of these items has an energy storage unit

FIGURE 2.76 Radiometer.

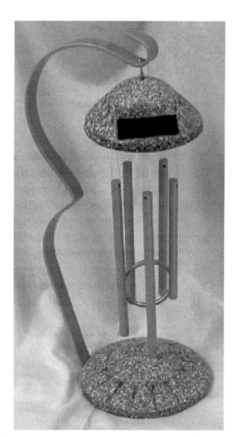

FIGURE 2.77 Sun chime.

FIGURE 2.78 Solar flashlight.

(battery) that allows it to operate when sunlight is not available. The common theme here is that the solar panel is essentially a battery charger and the device operates off of the stored energy. Coincidentally, solar battery chargers are also available. Some examples are shown in Figures 79 through 81.

Low-level power generation

Lightweight panels and generators are available that can be used to power nonsolar products where conventional electricity is not available. These panels can be used to power cell phones, computers, drills, and most other electrical devices we use every day (Figure 2.82). Solar panels can be purchased in many different sizes and some companies allow for parallel connections so that more than one panel can be used together for a greater power output.

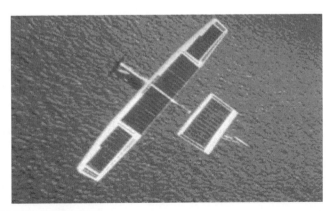

FIGURE 2.79 PV-powered plane.

FIGURE 2.80 PV-powered street light at UML.

FIGURE 2.81 Solar stainless pathway garden light.

2.8.2 Portable panels and generator

A more versatile product is the solar generator. This is essentially a device that consists of a solar panel, a battery, a regulator, and outlets. They come in a variety of sizes from 5 W to 750 W and up to a 1500 W surge. Solar Dynamics is one of the leading companies that makes these products. They also have available accessories

FIGURE 2.82 Portable solar panel.

such as a message board, flood lights, and extra panels. The outputs can be 3, 6, 9, and 12 V DC as well as AC. Solar Dynamic's large generator, the Harvester, is currently being used in some third-world countries to produce clean drinking water. An example of portable generators is displayed in Figure 2.83. The water purification system is shown in Figure 2.84.

FIGURE 2.83 Portable solar generator (DC plus 600 W AC).

FIGURE 2.84 PV-powered water purification unit.

2.8.3 PV-powered homes and remote villages

As stated earlier, when thinking of supplying their homes with solar power, most people think of the large unsightly solar panels that were popular in the early 1980s. Today, those panels are pretty much history and the newer technologies are form-fit, sleek, and attractive. They can supply a lot of power, have a long lifetime, and offer significant tax incentives. In addition, the utility companies are required by law to buy the unused power.

Figure 2.85 shows a PV-powered house in Massachusetts. PV homes need to have accurate power regulation and battery storage, as well as an inverter to convert the DC voltage to AC. Battery storage is typically an array of deep-cycle lead acid batteries. The number of batteries needed will vary depending on the average amount of electricity consumed each day and the type of solar panels used. In addition, Figure 2.86 shows PV-powered attic ventilation.

Solar shingles are used now in home construction, which reduces the cost of the PV system installed on houses (Figure 2.87). PV is also used to power remote villages and islands. Figure 2.88 shows a PV-powered Indian Reservation and Figure 2.89 shows a PV-powered village in Saudi Arabia. Table 2.2 contains a summary of stand-alone PV system applications.

FIGURE 2.85 PV-powered house in Massachusetts.

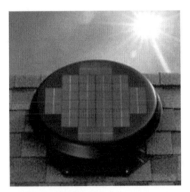

FIGURE 2.86 Solar-powered attic ventilation.

Source: Photo Courtesy of Solatube International

FIGURE 2.87 Installed solar shingles.

FIGURE 2.88 PV-powered Navajo Tribe Reservation.

FIGURE 2.89 PV-powered village in Saudi Arabia.

Table 2.2 Stand-alone Photovoltaic System Applications Summary

Battery charging
Lighting
  Lighting tunnel
  Street lights
Telecommunications
  Radar repeaters
  Microwave transmitters
  Portable military communication units

Table 2.2 Stand-alone Photovoltaic System Applications Summary—cont'd

Navigation Aids
  Beacons
  Buoys
  Lighthouses
Water Pumping
  Irrigation
  Groundwater wells
Cathodic (corrosion) protection
  Transmission towers
  Bridges
  Pipelines
  Wells
Solar domestic hot water system circulation pumps
Refrigeration of medical supplies and food
Environmental sensors
  Radiation samplers 0 Noise monitors
  Weather stations
Intrusion detectors/inhibitors
  Entry monitors
  Electric fences
Aviation aids
  Radar beacons
  Anemometers
Space conditioning
  Small coolers
  Ventilators
Water purification and desalination
Residential electricity
  Remote cabins, cottages, and vacation homes
  Larger systems with battery storage
  Hybrid systems (with wind or diesel power)
Remote traffic aids
  Highway signs
  Railroad crossings
  Traffic signals and counters
  Road ice sensors
Village and rural community power
Consumer products
  Calculators
  Watches and clocks
  Portable radios and television

FIGURE 2.90 A 12 MW PV farm in Arsstein, Germany.

2.8.4 Central power stations

One more application of a PV system is the cenral generation of electricity at a very large scale. One of the biggest such cental power stations is shown in Figure 2.90.

2.8.5 Future expectations

As technology continues to improve, companies will be able to offer their solar products at lower prices. Currently, PV systems are not inexpensive; they can cost as much as $7 to10 a watt as they call it key. Using these price estimates, the average home could save $300 to $600 per year. Furthermore, the life expectancy of the batteries is 15 to 20 years and the electronics should last much longer. The life of a quality solar panel is more than 25 years and for a solar shingle it is about 15 years. Also, since these systems are virtually maintenance free and do not have moving parts, they do not depreciate.

This combined with the need for more power generation will propel PV systems to commonplace. It is expected that more products will emerge for private as well as industrial use. It is unrealistic to expect that PVs will be used as our main source of power, at least in the near future. However, it certainly is proving itself to be an effective supplementary power source and there is plenty of room for growth. There are great advantages for continuing research and promoting this renewable energy source. The production of PV systems is increasing at a very high rate.

As can be seen from Figure 2.91, the future for PV use is very bright. This is because the cost is constantly going down due to mass production, innovation, and the invention of new products, such as rollable PV panels shown in Figure 2.92, and PV panels employing nanoparticles.

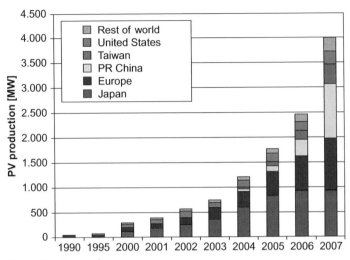

FIGURE 2.91 Applications of PV.

Source: PV News [Pvn 2008], Photon International [Pho 2008] and own analysis

FIGURE 2.92 Flexible PV sheets/Konarka.

2.9 UTILITY INTERCONNECTION ISSUES AND REQUIREMENTS

2.9.1 Introduction

The application of renewable energy sources (RES) has an impact on the utility companies, manifested in:

• The reduction in the demand side, which is equivalent to increasing the capacity of the utility company.

- The reduction in summer peaks load.
- The interaction with customers who used this RES, which includes:
  - Protect the safety and integrity of the utility system, islanding, and the contribution to short-circuit current in the utility system
  - Net Metering Rules, or buy back policy
  - Cost of interconnection with the qualified customer
  - Power quality, THD $< 5\%$; any individual harmonic is $< 3\%$
  - Acceptable power factor; customers with PF < 0.85 should pay penalty
  - Standards (e.g., level of noise)
  - Incentives (renewable energy credit, carbon credits)
- Large-scale use of RES needs diversity; it requires large load balancing areas in tandem with better regional planning.
- The overhead lines needed to carry the RES from abundant sources of renewable energy to customers is an important factor in their widespread use.
- Legislative issues: The zoning policy of installing small-scale RES in towns and cities has to be dealt with and legislated.
- Workforce: An abundant and skilled workforce to design, build, operate, and maintain RES. Support RES curriculum at universities and research institutions.

2.9.2 Interaction of the utility companies with RES customers

Because of the intermittent nature of the RES, their penetration level in a utility grid is important and needs to be figured out in light of the energy storage available to utility companies.

Public utility regulatory policies act
What does PURPA do?

The Public Utility Regulatory Policies Act (PURPA) requires electric utilities to purchase electric power from and sell electric power to cogeneration and small power production facilities. It exempts certain qualifying facilities from federal and state regulation as an electric utility. Each state regulatory authority and each nonregulated utility is responsible for implementation of the program according to the rules prescribed by the Federal Energy Regulatory Commission (FERC).

Who qualifies?

PURPA and the rules adopted by FERC define the facilities that will be considered for qualification:

- A small power production facility means a facility that produces electric energy solely by the use, as a primary energy source (greater than 75%), of biomass, waste, renewable resources, or any combination thereof; and has a power-production capacity that, together with any other facilities located within one mile using the same resource and owned by the same person, is not greater than 80 MW.

- A cogeneration facility means a facility that produces electric energy and steam, or forms of useful energy, which are used for industrial, commercial, heating, or cooling purposes.

Under PURPA, an electric utility holding company, or a subsidiary of either, may not maintain greater than a 50% interest in a qualifying facility. The exception to this rule is where the utility is a subsidiary of a company exempt from or declared not to be an electric utility under the Public Utility Holding Company Act.

At what rates will power be purchased from qualifying facilities?

Electric utilities must purchase electric energy and capacity made available by qualifying facilities at a rate reflecting the cost that the purchasing utility can avoid as a result of obtaining energy and capacity from these sources. State regulatory authorities are responsible for establishing the rates.

These *avoided costs* are defined as the costs to an electric utility of energy or capacity, or both, which the electric utility would have had to generate or construct itself or purchase from another source. This includes fuel costs and some operating and maintenance expenses as well as the capital costs of facilities needed to provide the capability to deliver energy. However, the rate provisions apply only if a qualifying facility chooses to avail itself of them. Negotiated agreements between qualifying facilities and electric utilities with conditions that differ from the PURPA rules are acceptable.

Each qualifying facility must bear the expense for any interconnection costs. These costs will be determined on a nondiscriminatory basis with respect to other customers with similar load characteristics. The interconnection costs of a facility that is already interconnected with a utility for purposes of sales are limited to any additional expenses incurred by the utility to permit purchases. The state regulatory authority is responsible for setting the amount and method of payment of costs, which may include reimbursement to the utility over a reasonable period of time.

At which rates will utilities sell electricity to qualifying facilities?

The provision requiring utilities to sell electricity to qualifying facilities further requires that the rates charged by the utilities be just and reasonable, in the public interest, and not discriminatory against the facilities in comparison to rates for sales to other customers served by the electric utility.

Utility interconnection issues and requirements

In the past little effort has been devoted to understanding and resolving the technical issues of integrating dispersed power-production systems into utility grids. Many utilities have, or are in the process of developing, requirements for interconnecting dispersed power-production facilities. However, the requirements placed on dispersed facilities are not uniform, varying from few restrictions to extremely complex or prohibitive requirements.

Reasonable interconnection requirements, which can ensure the safety and integrity of the utility system without placing an excessive penalty and expense on the customers desiring to interconnect, need to be developed. Unless such reasonable requirements are established for interconnecting dispersed power-production facilities to the utility grids, interconnection will not happen, and small residential or commercial PV systems will not significantly contribute to the nation's energy supply.

While PURPA requires utilities to interconnect any qualifying facility, it also requires the facility to pay for the cost of interconnection. The technical interconnection requirements imposed by utilities and their cost must be considered early in the development of a small power-production project. It is absolutely essential that owners, designers, and suppliers who are contemplating the installation of a utility-interactive, dispersed power-production facility be aware that the equivalent annual cost of the interconnection requirements may exceed the annual income from power production. It is in the best interest of all concerned to involve the utility early in the development of small power-production projects.

The key issue in utility-interactive PV systems is that they put electricity back into the system. Because the utility is ultimately responsible for the power system, a request for the interconnection of an independent power-production facility raises several technical questions and operational concerns—safety, power quality, power system protection, and metering. Essentially dispersed power-production facilities must (1) be safe for the utility customers and linemen, (2) not significantly degrade power quality on the system, (3) have protection equipment to respond to abnormal conditions, and (4) be metered to measure the power generated.

Safety. The questions and concerns involving safety issues are:

- How can utilities be assured that distribution lines will not be accidentally energized by independent power producers?
- How reliable are isolation devices provided by the independent producer?
- What additional devices and equipment should be provided?

On safety issues, the utility must be conservative. It is the utility that is ultimately responsible for the electrical transmission and distribution system.

Situations might arise in which a fuse or recloser isolates a section of line on which some independent generators are operating. This combination of line capacitance and load could possibly cause the generators to keep operating, resulting in a safety hazard to line crews working on the distribution system. Work rules may require manual disconnect and visual verification by the utility or live-line work before line crews can work on deenergized lines. There must be a means of disconnecting under utility control to isolate the independent power producer from the system.

Power quality. Interconnection of uncontrolled power producers raises several questions regarding the quality of power.

- Will the power supplied by the independent producer degrade the utility system quality? Specifically, will power quality variations cause problems for other utility customers?
- If interference problems result from operation of an independent producer, who would be available to provide assistance and at whose expense?
- Should these independent producers be required to meet specific wave-shape and noise objectives?

Small power production facilities should be designed to maintain power system quality. Power quality is evaluated in terms of voltage, frequency, harmonic distortion, power factor, and flicker.

Utilities provide power within fixed ranges of voltage and frequency so electric appliances and other equipment will not be damaged. The typical 120 to 240 V, single-phase power provided to residences is within 114–125/228–252 V in accordance with Standard System Voltages (recommended by the American National Standards Institute: ANSI C84.l). It is reasonable to expect dispersed generation facilities to provide power within these same limits.

An additional power quality problem is flicker due to voltage variations. Flicker causes visible flickering of incandescent lamps and shrinking and expanding pictures on television screens. It is caused by the varying output of power producers. The exact level where flicker becomes objectionable is difficult to establish. The greatest effects would be on the premises of the production facility or others served by the same transformer. The utility may require the producer to provide suitable equipment to limit flicker or use a dedicated transformer to limit the effects on other customers.

Alternating current equipment and power systems need a relatively smooth, undistorted waveform. Distortions or harmonics are voltages or currents at frequencies that are multiples of the 60-Hz fundamental frequency. Excessive levels of harmonic voltages and currents can cause excessive heating of motors, transformers, and capacitors as well as communications interference. Harmonic content must be limited to avoid interference and damage to equipment and other customers.

The inverter or power-conditioning system used on photovoltaic power-production facilities should be designed to minimize harmonic generation. If objectionable harmonics are produced by independent power producers, the utility must require that filters or other equipment be installed by the producer to eliminate problems. Filtering, if required, is likely to be costly.

Some power-conditioning systems consume a significant amount of reactive power (VAR), decreasing the power factor on the distribution system feeder. Power factor is defined as the ratio between true power and apparent power. A low power factor increases the amount of power that must be generated and increases the losses in a utility distribution system.

Most customers are billed for true power used without metering their power factor or reactive power. However, the use of special metering for large industrial customers to monitor the power factor is standard practice. Customers having power factors below 0.85 pay a penalty for the power factor. Independent power systems with poor power factors may be required to use specialized metering and/ or install capacitor banks to improve the power factor; either alternative could be costly.

Power system protection. The introduction of independent power producers raises questions about equipment and line protection.

- How can utilities be assured that power system equipment will be protected against damage from normal and abnormal operation of the independent producer?
- Will feeder line sectionalizing devices operate correctly with independent generators operating in parallel?

The power producer must be disconnected when faults or other abnormal conditions occur on the utility system.

Protection equipment that interrupts short circuits on the customer side is required by code. However, the effect of shorts on the customer and the associated protection equipment may disrupt the coordination of fuses and other sectionalizing devices in the distribution system. Power system protection is designed to minimize service interruptions to customers while isolating the problem. Fuses, reclosers, and relay-operated feeder circuit breakers that respond to current are all coordinated to ensure that the equipment nearest to the fault (between the fault and the substation) will trip first, isolating a minimum number of customers.

Fuses or other devices under the control of utilities may be required at the independent power-production facility to maintain system coordination. Line sectionalizing schemes must also be reevaluated if various independent power producers are interconnected. Small systems are not likely to contribute significantly to fault currents. However, if large systems, or a large number of small producers, are added in a specific locality, problems may arise.

Metering. Metering practices are set by the contract terms between the producer and utility on an individual basis. Elaborate metering resulting from potential problems with power quality and type of rates (on peak and off peak) can be costly for a small powerproducer. Net energy metering with one watt-hour meter can be used if the purchase and sales rates are equal.

Since most meters are calibrated to measure power in one direction, the simplest metering scheme for small power producers is to use two meters, one to measure the kWh produced and one to measure the kWh consumed. The metering scheme becomes more complicated and costly when complex rates or power factor problems exist.

Generating capacity and other issues

The nation benefits by becoming less vulnerable to interruption of finite, nonrenewable energy sources. Essentially, photovoltaic systems are a technology that should be pursued. There are, however, institutional, economic, and technical concerns that must be resolved to ensure that these benefits can be realized. Progress is being made but technical issues, in particular, require further evaluation. The safety hazards caused by isolated operation of small power-production facilities are the greatest current concern.

More experience is also required in the technical areas of harmonics, power factor correction, and protective equipment furnished by power conditioning system manufacturers. Tests and experiments conducted in cooperation with utilities, such as the FSEC's DOE Photovoltaic Southeast Residential Experiment Station, should provide some of the answers needed to develop reasonable, economical interconnection requirements for photovoltaic systems.

The technical issues involved in interconnecting PV power-production facilities can be resolved if owners, designers, and suppliers work with utilities on the unique problems of interconnection. As photovoltaic systems are "hooked up" to the grid, and interfacing and protection equipment becomes reliable, "reasonable" requirements will be promulgated by utilities. Only then can photovoltaic application take its place as a reliable alternative power-distribution system.

PROBLEMS

2.1 A photovoltaic module has the following parameters: $I_{sc} = 4.42$ A, $V_{oc} = 18.1$ V, $V_{mp} = 13.6$ V, $I_{mp} = 3.46$ A recorded at an irradiance of $S = 1000$ W/m^2. The length of the module is $L = 1.2$ m and the width of the module is $W = 0.6$ m.

   a) Calculate the fill factor, FF.

   b) The module is connected to a variable resistive load at an irradiance of $S = 900$ W/m^2. The voltage across the resistance was $V = 15.2$ V and the current flowing in the resistance was $I = 2.86$ A. Calculate the module efficiency, η.

   c) The irradiance is increased from $S = 900$ W/m^2 to $S = 1000$ W/m^2. How much would you increase or decrease the load resistance to extract maximum power from the photovoltaic module?

2.2 A PV module has the following parameters recorded at an irradiance of $S = 400$ W/m^2.

$$I_{sc} = 1.1 \text{ A}, \ V_{oc} = 18.68 \text{ V}, \ V_{mp} = 14.75 \text{ V}, \ I_{mp} = 0.91 \text{ A}$$

   a) Calculate the fill factor.

   b) What should be the resistance to be connected across the module to force it to deliver maximum power to the resistance?

2.3 Two dissimilar PV modules are subjected to the same irradiances and temperatures.

   • The short-circuit current output of the first module is $I_1 = 4$ A and the open-circuit voltage output is $V_1 = 17.55$ V.

- The short-circuit current output of the second module is $I_2 = 2.3$ A and the open-circuit voltage output is $V_2 = 18.35$ V
  - **a)** The two PV modules are connected in series, as in Figure 2.35. Calculate the combined short-current output, I, and the combined open-circuit voltage, V, across both circuits.
  - **b)** The two modules are connected in parallel, similar to Figure 2.40. Calculate the combined short-circuit current, I and Voltage V.

2.4 Two similar PV modules are subjected to the same irradiances and temperatures. The short-circuit current output of the first module is $I_1 = 2.3$ A and the open-circuit voltage across the module terminal is $V_1 = 18.35$ V. The short-circuit current output of the second module is $I_2 = 2.3$ A and the open-circuit voltage across its terminals is $V_2 = 18.35$ V.
  - **a)** The two modules are connected in series, similar to Figure 2.33. Find the combined short-circuit current, I, and the combined open-circuit voltage, V.
  - **b)** The two modules are connected in parallel, similar to Figure 2.38. Calculate the combined short-circuit current, I, and the combined open-circuit voltage, V.

2.5 A house has the appliances listed in the following tables along with their daily energy consumption.

Appliance	Monthly consumption in kWh
Coffee maker	10
Microwave oven	9
Oven	42
Refrigerator	125
Dryer	78
Hand iron	5
Lighting	50
Spotlight	45
Clock	1.5
Vacuum cleaner	4
Water heater	195
Hair dryer	2
Television	7
Radio	54
Computer	2
Electric pan	8
Blender	12

Appliance	Rated wattage, W	Daily hours used
Electric piano	30	3
Mixer	15	1
Amplifier	30	3
Synthesizer	5	5
Bass	8.4	3

a) Calculate the house's daily energy demand.

b) If the battery voltage is 12 V, calculate the house's daily amp-hours demand.

c) If the available battery has 12 V, a capacity of 300 AH, 80% round-trip efficiency, and 80% depth of discharge, calculate the number of batteries needed for 7 days of storage.

d) The available PV module has 120 W$_p$; the house's location has 5 peak sun hours. Calculate the number of modules needed to satisfy the energy demand of the house.

2.6 If the integrated area under the curve A1 in Figure 2.3 is 4.5 kWh, calculate the number of peak sun hours at this location.

2.7 The voltage output of an inverter used as part of the PV system is similar to what is shown in Figure 2.71. If the third harmonic is 40% of the fundamental, the fifth harmonic is 50% of the fundamental, and the seventh harmonic is 20% of the fundamental:

a) Calculate the total harmonic distortion.

b) Is that inverter acceptable to IEEE? Why?

2.8 The output current waveform of an inverter that is used in a photovoltaic grid-interactive system contains the following harmonics: the third harmonic is 40% of the fundamental, the fifth harmonic is 50% of the fundamental, and the seventh harmonic is 20% of the fundamental.

a) Calculate the THD of the current waveform.

b) Is that acceptable by IEEE standards? Why?

2.9 The I–V characteristics of a photovoltaic module is given in Figure 2.93.

a) Use the figure to calculate the PV series resistance, R_s in ohms.

b) Use the figure to calculate the shunt resistance, R_{sh} in ohms.

2.10 If a load resistance of 25 ohms is connected across the PV module of Figure 2.93, then:

a) Calculate the voltage across the PV module.

b) Calculate the current flowing in the load resistance.

c) Calculate the power output of the module.

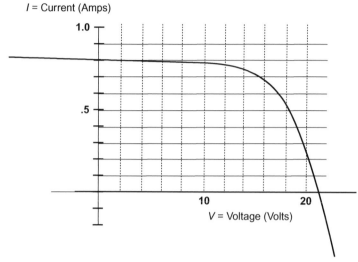

I = Current (Amps)

V = Voltage (Volts)

FIGURE 2.93 The I–V characteristics of a PV module.

References

[1] Lynch WA, Salameh ZM. Simple electo-optically controlled dual-axis sun tracker. Sol Energy 1990;45(2):65–9.

[2] Salameh Z, Mulpur AK, Dagher F. Two-stage electrical array reconfiguration controller for PV-powered water pump. Sol Energy 1990;44(1):51–6.

[3] Salameh Z, Taylor D. Step-up maximum power point tracker for photovoltaic arrays. Sol Energy 1990;44(1):57–61.

[4] http://en.wikipedia.org/wiki/P-n_junction.

[5] *http://en.wikipedia.org/wiki/File:Solar_Spectrum.png; [accessed 10.01.12].

[6] NASA. Science photo library, www.sciencephoto.com/media/337235/enlarge; [accessed 10.01.12].

[7] *http://en.wikipedia.org/wiki/Photovoltaics.

[8] Stand alone PV system sizing worksheet (example), *dnr.louisiana.gov/sec/execdiv/techasmt/energy_sources/renewable/PV%20Sizing%20Guide%20w- blank.pdf.

[9] www.yardbright.com/files/product_336_1_big.jpg; [accessed 10.01.12].

[10] www.solar-powered-garden-lights.net/Portable-Solar-Panels.html; [accessed 10.01.12].

[11] www.uni-solar.com/2010/11/uni-solar-banks-on-efficiency-cost-bos-improvements-to-weather-the-storm/; [accessed 10.01.12].

[12] www.cheapdiysolarpanels.net/solar-panel-types/solar-powered-generator/;   [accessed 10.01.12].

[13] http://safedrinkingwater.net.au/waterpurifierfilter/sunrunner; [accessed 10.01.12].

[14] http://homebuilding.thefuntimesguide.com/2008/05/solar_powered_attic_vent_fans.php; [accessed 10.01.12].

[15] www.westernsun.org/WebServer/Documents/Unisolar/unisolar%20shakes.html.

[16] www.nrel.gov/clean_energy/photovoltaic.html.

[17] http://science.nasa.gov/headlines/y2002/solarcells.htm.

[18] http://science.nasa.gov/headlines/y2002/08jan_sunshine.htm.

[19] *http://ens-newswire.com/2010/01/12/state-of-the-world-2008-environmental-woes-sow-seeds-of-sustainability (accessed 10.03.12).

[20] www.nanosolar.com; [accessed 10.03.12].

[21] wwwrewci.com/soatvefan.html; [accessed 10.03.12].

[22] www.pvresources.com/en/top50pv.php; [accessed 10.03.12].

[23] http://geneva.usmission.gov/2012/04/23/u-s-supports-saudi-arabia%E2%80%99s-clean-energy-goals/; [accessed 10.03.12].

[24] Nuclear News. Solar energy: a new economy for Navajo tribes, http://nuclear-news.net/2012/04/27/solar-energy-a-new-economy-for-navajo-tribes/; [accessed 10.04.12].

[25] www.konarka.com/index.php/company/tech-sheets-and-brochures/; [accessed 10.04.12].

[26] www.solardesign.com/projects/project_summary.php?title=Residential%20Design&type=residential; [accessed 10.04.12].

[27] http://ihome21.kennesaw.edu/new/img/solar/solarchallenger.jpg; [accessed 10.04.12].

[28] http://keetsa.com/blog/eco-friendly/bogo-solar-flashlight/; [accessed 10.04.12].

[29] http://images.shopcasio.com/imagesEdp/p127669b.jpg; [accessed 10.04.12].

[30] www.kaboodle.com/hi/img/2/0/0/165/9/AAAAAlH4UWEAAAAAAWWQrw.png; [accessed 10.04.12].

[31] www.starmajic.com/Solar-Radiometer.html; [accessed 10.04.12].

[32] www.tomtop.com/media/catalog/product/cache/1/image/ced77cb19565515451b3578a3bc0ea5e/2/0/20101214_img_010370.jpg; [accessed 10.04.12].

[33] Hershey Energy Systems, www.hersheyenergy.com/images/Harmonics_Graph.jpg, [accessed 10.04.12].

[34] Borowy B, Salameh Z. Methodology for the optimally sizing the combination of a battery bank and PV array in a wind/PV hybrid system. IEEE Trans Energy Conversion 1996;11 (2):367–75.

[35] www.solarserver.com/solarmagazin/solar-report_0509_e.html; [accessed 10.15.12].

[36] http://peakoildebunked.blogspot.com/2008/12/387-world-photovoltaic-pv-production.html; [accessed 10.15.12].

[37] www.e-education.psu.edu/egee401/content/p7_p5.html; [accessed 10.15.12].

[38] www.waoline.com/science/newenergy/Photovolt/SolarConcentrators.htm; [accessed 10.15.12].

[39] http://sunxran.files.wordpress.com/2009/08/solarpowermap.gif; [accessed 10.15.12].

Wind Energy Conversion Systems

3.1 INTRODUCTION

Wind energy conversion systems (WECS) are designed to convert the energy of wind movement into mechanical power. With wind turbine generators, this mechanical energy is converted into electricity and in windmills this energy is used to do work such as pumping water, mill grains, or drive machinery. The first wind machines were probably vertical axis windmills used for grinding grain in Persia dating back to 200 BC (Figure 3.1). They had a number of arms on which sails were mounted, with the sails initially made from bundles of reeds.

During the tenth century, horizontal axis-mounted windmills first appeared in the Mediterranean region. These windmills were fixed permanently to face the prevailing coastal winds. Several hundred years later in Europe, horizontal windmills were operated with a manual mechanism that rotated the whole windmill to face the wind. These were used for grinding grains and pumping water.

Since earliest recorded history, wind power has been used to move ships, grind grain, and pump water. There is evidence that wind energy was propelled boats along the Nile River as early as 5000 BC. Within several centuries before Christ, simple windmills were used in China to pump water.

In the United States, millions of windmills were erected as the American West was developed during the late nineteenth century (Figure 3.2). Most of them were used to pump water for farms and ranches. By 1900, small electric wind systems were developed to generate direct current, but most of these units fell into disuse as inexpensive grid power was extended to rural areas during the 1930s. By 1910, wind turbine generators were producing electricity in many European countries.

All renewable energy (except tidal and geothermal power), and even the energy in fossil fuels, ultimately comes from the Sun, which radiates 100,000,000,000,000 kilowatt hours (kWh) of energy to the Earth per hour. In other words, the Earth receives 10^{18} watts (W) of power. About 1 to 2% of the energy coming from the Sun is converted into wind energy. That is about 50 to 100 times more than the energy converted into biomass by all plants on Earth.

As long as the Sun is heating the Earth, there will always be winds because temperature differences drive air circulation. The wind blows because the heating rates of the Earth differ, therefore, as the rate of evaporation of air over one area is different than another, there is a pressure differential (Figure 3.3). This causes the

FIGURE 3.1 A Persian windmill.

FIGURE 3.2 A lift-type American windmill.

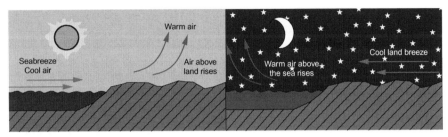

FIGURE 3.3 The coastal wind flow.

higher pressure from one area to flow to another. The satellite picture taken by NOAA shows different temperatures in different spots of the Earth—blue indicates the coolest and red indicates the warmest (Figure 3.4).

Near the world's bodies of water, the cool air over the water flows to the land. This is reversed during the night, when the cool air over the quickly cooled land flows toward the water, where the air over the land is less dense because water retains the Sun's heat longer.

Wind energy is a commercially available renewable energy source, with state-of-the-art wind plants producing electricity at about $0.05 per kWh. However, even at that production cost, wind-generated electricity is not yet fully cost-competitive with coal- or natural-gas-produced electricity for the bulk the market.

The wind is a proven energy source; it is not resource-limited in the United States, and there are no insolvable technical constraints. There are a lot of methods that describe current and historical technology, characterize existing trends, and describe the research and development required to reduce the cost of wind-generated electricity to full competitiveness with fossil-fuel-generated electricity for the bulk electricity market. Such potential markets can be described.

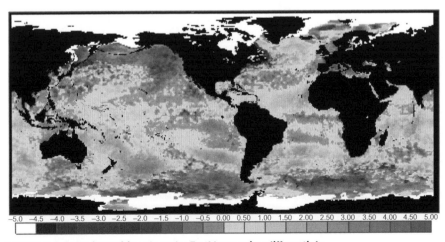

FIGURE 3.4 Color intensities show the Earth's warming differentials.

Winds arise because of the uneven heating of the Earth's surface by the Sun. One way to characterize winds is to use seven classes according to power density: class 1 is the lowest and class 7 is the greatest (shown in Figure 3.5). The wind power density is proportional to the wind velocity raised to the third power (velocity cubed). For utility applications, class 4 or higher energy classes are usually required.

Class 4 winds have an average power density in the range of 320 to 400 W/m^2, which corresponds to a moderate speed of about 5.8 m/s (i.e., 13 mph measured at a height of 10 m). Researchers estimate that there is enough wind potential in the United States to displace at least 45 quads of primary energy annually used to generate electricity. This is based on "class 4" or greater winds and the judicious use of land. For reference, the United States used about 30 quads of primary energy to generate electricity in 1993.

Denmark was the first country to use wind turbine generator to generate electricity in 1890 (Figure 3.6). The first modern US wind-turbine generator was erected and put into service in 1941 in Rutland, Vermont; it was called Grandpa's Knob. The turbine had diameter of 55 m and was rated at 1.25 MW (megawatts) at a speed of 13.5 m/s. It was operated for 18 months before the bearings failed.

Large-scale, grid-connected wind energy installations used for generating electricity have made enormous strides over the last 15 years. By the end of 1996, the United States hosted approximately 1750 MW of wind energy-generating capacity;

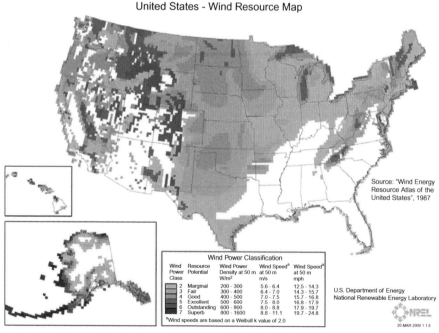

FIGURE 3.5 Wind class definitions and areas in the United States.

FIGURE 3.6 A Dutch-type windmill.

this compares with more than 4500 MW of capacity operated in Europe, India, and other locations. Capital cost, reliability, and energy conversion efficiency have increased to the point where these renewable energy systems can compete economically under many circumstances with conventional generation technologies such as nuclear and modern coal-fired plants.

The installed capital costs of wind-driven generating systems decreased from more than $2500 per kilowatt (kW) in the early 1980s, to $1000 per kW or less for large-scale installations in the mid-1990s. The costs of unscheduled and preventive maintenance also decreased in the same time period, from more than $0.05 to less than $0.01 per kWh.

These improvements have reduced the level cost of wind energy systems from more than $0.15 to less than $0.05 per kWh—not including the federal $0.02.1-kWh tax credit now available. Design and manufacturing advances, the further result of ongoing research and development programs, and the realization of large production volumes promise to reduce these costs still further to the range of $0.02.5 to $0.03.5 per kWh over the next 10 years.

Meanwhile, improvements in rotor aerodynamics and turbine-operating modes, along with increases in turbine size, have boosted the conversion efficiency of wind energy systems. Under good wind conditions, modern systems typically achieve capacity factors of 28% or more.

Detailed knowledge about the wind is essential if the design and economics of large windmills are to be properly understood and evaluated. Generally speaking, the highest wind speed sites are on exposed hilltops, offshore, or on coastal sites. Because of the difference in terrain, wind characteristics may be widely different between these locations. Information is required with regard to various parameters, both in general terms for each type of site and in detail for specific sites.

Such data include mean wind speed, the distribution about the mean wind speed, directional data, variations in wind speed in the short term (gusting), daily and annual or seasonal variations, and the changes of wind speed and direction with height. Although records of wind speed have been kept at some places for very long periods of time and meteorological stations recording wind speed and direction are very widely distributed over an area like the United Kingdom, it is perhaps surprising that there are still many gaps in our knowledge about the detailed behavior of the wind. This is particularly true for wind behavior over the open waters.

3.2 THE CHARACTERISTICS OF WIND

Wind speed is the most important factor influencing the amount of energy a wind turbine can convert to electricity. Increasing wind velocity increases the amount of air mass passing the rotors; so increasing speed will also have an affect on the power output of the wind system. The energy content of the wind varies with the cube (the third power) of the average wind speed.

3.2.1 Average wind speed (mean)

The best way of measuring wind speed at a prospective wind turbine site is to fit an anemometer at the top of a mast that has the same height as the expected hub of the wind turbine to be used. This way one avoids the uncertainty involved in recalculating wind speeds to a different height. By fitting the anemometer to the top of the mast, one minimizes the disturbances of airflows from the mast itself. If anemometers are placed on the side of the mast, it is essential to place them toward the prevailing wind direction to minimize the wind shade from the tower.

3.2.2 Wind speed–duration curve and height variations

This curve in Figure 3.7 shows how many hours a year the wind speed is more than a certain value.

The combined effects of the pressure gradient and the Earth's rotations largely determine the character of the high-altitude free stream or geotropic wind. In the first

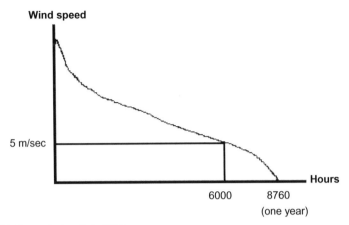

FIGURE 3.7 A speed–duration curve.

few hundred meters above the Earth's surface, however, there exists a turbulent layer known as the *boundary* or mixing layer. Within this layer, wind speed and direction vary with height due to surface frictions and temperature gradients. The variation depends on the type of terrain, atmospheric stability, and the distance over which the wind has already traveled. The latter parameter determines whether steady-state conditions have been achieved.

Even the largest wind turbines operate within the boundary layer, and it is therefore important to understand how the wind changes over the vertical scale of the machine and its supporting structure. The wind speed, in general, increases with height above the Earth's surface and empirically it has been found that a power law gives a good fit over a limited height range. The simplest relationship between the velocity, call it V_h at some height h, and the measured velocity, call it V_o, at some reference height, h_o, is given by the equation:

$$\frac{V_h}{V_o} = \left(\frac{h}{h_o}\right)^{\alpha} \tag{3.1}$$

where the exponent α depends on the roughness of the surface. For open land, α is frequently taken to be about 0.14, whereas for a calm sea, α may be as low as 0.1. A plot of wind speed with height for these values of α is shown in Figure 3.8. The nature of the surface also influences the direction in the boundary layer.

It is a matter of common experience that the wind rarely flows with a constant velocity close to the surface. Because of the low viscosity of air, it tends to become turbulent at low velocities leading to the phenomena of gusts and lulls. These are associated with the vertical motion of eddies, bringing down faster moving air from above in the case of gusts. This vertical motion is, of course, essential for maintaining a surface wind against the tendency of the frictional force to reduce the airflow close to the surface.

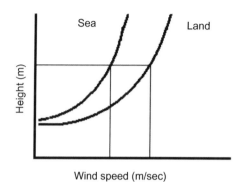

FIGURE 3.8 The wind speed as a function of height and terrain.

The occurrence and behavior of gusts are important to the windmill designer in two respects. First, they must be taken into account in evaluating the stresses to which the rotor and supporting structure will be subjected, and second, they will influence the design of the logic used to determine machine orientation, pitch control, and consequently power output. In the case of clusters or arrays of windmills, the spatial correlation, as well as the temporal variation of gusts, is particularly important.

EXAMPLE 3.1
The wind speed at the height of 10 m is 5 m/sec. Calculate the wind speed at a height of 40 m.

a) The location is on land, $\alpha = 0.14$

$$V_h = v_o \left(\frac{h}{h_o}\right)^\alpha = 5 \left(\frac{40}{10}\right)^{0.14}, V_h = 6.07 \text{ m}$$

b) The location is on the water, $\alpha = 0.1$

$$V_h = v_o \left(\frac{h}{h_o}\right)^\alpha = 5 \left(\frac{40}{10}\right)^{0.1}, V_h - 5.74$$

3.2.3 Wind speed variations with time and direction

In areas where weather patterns tend to be dominated by the passage of fronts over a time scale of a day or two, they determine the coarse structure of the wind. Within these general patterns, variations occur over periods of tens of minutes and these are very significant in terms of windmill performance. Changes over this time scale influences the amount of time a conventional generating plant may be needed to respond rapidly to changes in wind power output. Daily, seasonal, and annual variations also occur and can have a significant impact on wind energy economics.

On a daily basis, the pattern of wind speeds, and thus the energy available from the winds, is not entirely random. Above land, atmospheric heat loss is greatest in early afternoon and this is particularly pronounced in the summer months. This is reflected in both higher mean wind speeds and greater turbulence at this time. At sea, the position is much less clear; there are little diurnal variations in water surface temperature and an inverse relationship between onshore and offshore atmospheric stability is to be expected. However, this will be significantly affected by land and sea breeze effects, distance offshore, wave height, seasonal variations, and wind direction.

These diurnal changes in wind are important because there exists well-defined variations of demand for electricity during the day and this affects the value of the wind energy. The fact, for example, that demand peaks in late afternoon in the United Kingdom, means that the statistically significant extra wind energy produced at this time by windmills sited on land may slightly increase its value to the electricity supply system.

The wind direction shall be determined by averaging the direction over a 2-minute period. When the wind direction sensor(s) is out of service, at designated stations, the direction may be estimated by observing the wind cone or tee, movement of twigs, leaves, smoke, and so on, or by facing into the wind in an unsheltered area.

The wind direction may be considered variable if, during the 2-minute evaluation period, the wind speed is 6 knots or less. Also, the wind direction should be considered variable if, during the 2-minute period, it varies by 60 degrees or more when the average wind speed is greater than 6 knots.

The data on both wind speeds and wind directions from the anemometer(s) are collected on electronic chips on a small computer, a data logger, which may be battery-operated for a long period. Once a month or so we may need to go to the logger to collect the chips and replace them with blank ones for the next month's data. Wind speeds are usually measured as 10-minute averages in order to be compatible with most standard software. The results for wind speeds will be different if you use different periods for averaging, as we'll see later.

3.2.4 **Wind rose**

Strong winds usually come from a particular direction; to show the information about the distributions of wind speeds, and the frequency of the varying wind directions, we can draw a so-called *wind rose* on the basis of meteorological observations of wind speeds and wind directions. The picture in Figure 3.9 shows the wind rose for Brest on the Atlantic coast of France.

The compass has been divided into 12 sectors, one for each 30 degrees of the horizon. (A wind rose can also be drawn for 8 or 16 sectors, but 12 sectors tend to be the standard set by the European Wind Atlas from which this image was taken).

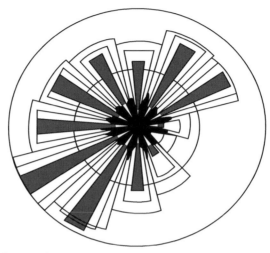

FIGURE 3.9 Graphic of a wind rose.

The radius of the 12 outermost, wide wedges gives the relative frequency of each of the 12 wind directions (i.e., how many percent of the time is the wind blowing from that direction). The second wedge gives the same information, but multiplied by the average wind speed in each direction.

The result is then normalized to add up to 100%. This tells us how much each sector contributes to the average wind speed at a particular location. The innermost (red) wedge gives the same information as the first, but it is multiplied by the cube of the wind speed in each location. The result is then normalized to add up to 100%. This tells us how much each sector contributes to the energy content of the wind at this particular location.

A wind rose gives information on the relative wind speeds in different directions (i.e., each of the three sets of data—frequency, mean wind speed, and mean cube of wind speed) has been multiplied by a number, which ensures that the largest wedge in the set exactly matches the radius of the outermost circle in the diagram. Wind roses vary from one location to the next; they actually are a form of meteorological fingerprint.

Wind roses from neighboring areas are often fairly similar, so in practice it may sometimes be safe to interpolate (take an average) of the wind roses from surrounding observations. The wind rose, once again, only tells us the relative distribution of wind directions, not the actual level of the mean wind speed.

3.2.5 **Wind shear and gusts**

High above ground level, at a height of about 1 kilometer, the wind is barely influenced by the surface of the Earth at all. In the lower layers of the atmosphere, however, wind speeds are affected by the friction against the surface of the Earth

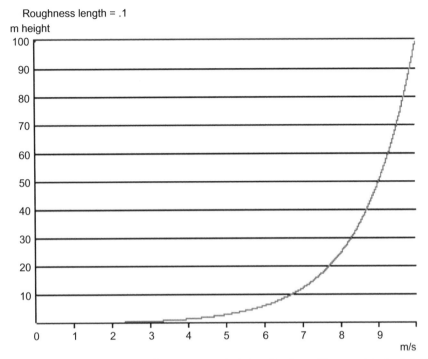

Roughness length = .1 m height

FIGURE 3.10 **The effects of the wind speed on wind shear/log-law with respect to the height.**

(Figure 3.10). In the wind industry, one distinguishes between the roughness of the terrain, the influence from obstacles, and the influence from the terrain contours, which is also called the *orography*. We will be dealing with orography when we investigate the so-called speed up effects, i.e.; however, the roughness will not fall neatly into any of the roughness classes, so a bit of averaging is necessary. We have to be very concerned with the roughness in the prevailing wind direction. In those directions, we look at a map to measure how far away there is unchanged roughness.

Gust is defined as the rapid increase in wind speed over a very short period of time; 20 sec is standard time. One must design the wind turbine structure to withstand peak gust conditions. The sudden change in wind speed poses a hazard to large wind turbines. Wind gusts also require a complex control system.

3.2.6 Wind speed distribution and probability functions

Wind speed is a variable quantity; a probability distribution of the wind speed provides information about the long-term characteristics of a site. The probability distribution is used to calculate the available wind energy output. To effectively use the wind data, it is important to understand the following terminologies:

Mean wind speed

The mean wind speed, V_{wm}, is the average of all observed wind speeds:

$$V_{wm} = \frac{\sum_{i=1}^{n} V_{wi}}{n} \tag{3.2}$$

where,

V_{wi} = observed wind speed
n = the number of observations

Variance Δ^2

The average of the deviation of the different speeds from the mean wind speed is called variance. It is defined by;

$$\delta^2 = \frac{1}{n-1} \sum_{i=1}^{n} [V_{wi} - V_{wm}]^2 \tag{3.3}$$

The standard deviation

The standard deviation, δ, is defined as the square root of the variance:

$$\delta = \sqrt{\text{variance}}$$

EXAMPLE 3.2

In a location five wind speeds were observed: 2, 4, 7, 8, and 9 m/sec. Calculate:

- V_{wm}, δ^2 and δ
- $V_{wm} = (2 + 4 + 7 + 8 + 9)/5 = 6$ m/sec
- $\delta^2 = (1/4)[(2 - 6)^2 + (4 - 6)^2 + (7 - 6)^2 + (8 - 6)^2 + (9 - 6)^2$
  $= 8.5$ m^{2}sec^2
- $\delta = \sqrt{8.5}$ 2.92 m/sec

P(V_{wi}): Probability that a wind speed, V_{wi}, is observed

$$P(V_{wi}) = m_i/n$$

where,

m = the number of observations of the speed V_{wi}
n = total number of observation

Cumulative distribution function

The probability that a measured wind speed, $F(V_{wj})$, is less or equal to V_{wi}:

$$F(V_{wj}) = \sum_{j=1}^{i} P(V_{wj})$$

EXAMPLE 3.3

On a site, the number of wind speed observations is $n = 211$; each wind speed, V_{wi}, is observed, m_i. Calculate $P(V_{wi})$ and $F(V_{wi})$, the results are shown in the following table.

i	V_{wi}, m/sec	m_i NO of observation	$P(V_{wi})$	$F(V_{wi})$
1	0	0	0	0
2	1	0	0	0
3	2	15	0.071	0.071
4	3	42	0.199	0.270
5	4	76	0.36	0.63
6	5	51	0.242	0.872
7	6	27	0.128	1.00

Probability density function

The probability density function, $f(V_w)$, is the probability that the wind speed is within a certain range.

$$f(V_w) = dF(V_w)/dV_w$$

The most common probability density functions are:

<u>Weibull</u>

$$f(V_w) = \frac{K}{C}\left(\frac{V_w}{C}\right)^{K-1} \exp\left[-\left(\frac{V_w}{C}\right)^K\right]$$

C : scale factor
K : shape factor   (3.4)

$$V_{wm} = C\Gamma\left(1 + \frac{1}{K}\right)$$

$$\sigma^2 = C^2\left\{\Gamma\left(1 + \frac{2}{K}\right) - \Gamma^2\left(1 + \frac{1}{K}\right)\right\}$$

The Weibull is a two-parameter distribution function, this makes it more versatile:

$$\underline{\underline{Rayliegh}}$$

$$f(V_w) = \left(\frac{V_w}{C^2}\right) \exp\left[\left(\frac{-V^2}{2C^2}\right)\right] \tag{3.5}$$

$$C : scale$$

The Rayliegh has only one parameter that makes it simpler to use.

$$\underline{\underline{Beta}}$$

$$f(V_w) = \left(\frac{V_w}{V_{w,max}}\right)^{a-1} \left(1 - \frac{V_w}{V_{w,max}}\right)^{b-1} \left[\frac{1}{B(a,b)V_{w,max}}\right]$$

$$B(a,b) = \frac{\Gamma(a)\Gamma(b)}{\Gamma(a,b)} \tag{3.6}$$

$$a = \frac{V_{wm}}{V_{w,max}} \left[\frac{V_{wm}(V_{w,max} - V_{wm})}{\delta^2}\right]$$

$$b = a\left[\frac{V_{w,max} - V_{wm}}{V_{wm}}\right]$$

where,

a and b = the parameters of the Beta distribution function
$B(a, b)$ = the β function
$\Gamma(a), \Gamma(b)$ = the γ function
V_{wm} = mean wind speed
$V_{w,max}$ = maximum wind speed.

3.3 THE AERODYNAMIC THEORY OF WINDMILLS

As with any device that uses the wind as a source of energy, a wind turbine has to follow the aerodynamic theories that apply to its use. Aerodynamics is a branch of mechanics that deals with the motion of air (and other gases) with the effects of such motion on bodies in the medium. The primary aerodynamic forces that act on a wind turbine are lift, drag, and stall, although the design must also contend with the problem of turbulence. *Lift* is the aerodynamic force having a direction perpendicular to the direction of motion. *Drag* is the aerodynamic force exerted on an airfoil, or other aerodynamic body, that tends to reduce its forward momentum. *Stall* is the loss of lift due to a change in the angle of attack. *Turbulence* is the haphazard secondary motion caused by eddies in the moving medium, in this case the airstream.

The terms lift and drag are sometimes used to describe the two basic layouts used in the design of wind turbines. A horizontal axis wind turbine, where the axis of rotation is parallel with the wind stream, is also referred to as a lift machine (refer to Figure 3.2). While there is also drag present in this design, lift is the primary force that drives the propeller. In a vertical axis machine, where the axis of rotation is perpendicular to the wind stream, the primary force used in generating the rotation is drag. Even within these two formats, there can be significant differences from one design to the next.

At the most basic level, a wind turbine operates according to Newton's third law; if object A exerts a force on object B, then object B exerts an oppositely directed force of equal magnitude on A. This is particularly evident, and intuitive, when applied to drag-type machines. In the vertical axis format, such as a Savonius rotor shown in Figure 3.11, it is self-evident as to how the blades are pushed downwind by the wind stream.

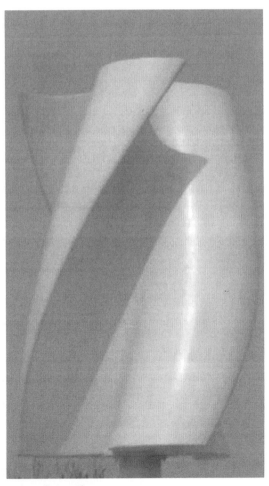

FIGURE 3.11 A Savonius wind turbine.

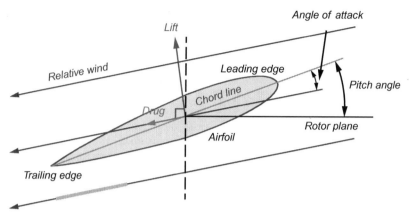

FIGURE 3.12 Forces and angle of attack on a propeller blade.

This simplicity is related to their relatively low efficiency. Drag-type machines tend to have an empirically measured efficiency of between 12 to 15%. While this seems like drag machines are useless, they do have other advantages that make up for their low efficiency. For example, they are exceedingly simple to build and maintain, exhibit a high start-up torque, and are omnidirectional. To make a wind machine more efficient, one must narrow the analysis of the machine, and the shape of an airfoil and the angle at which it strikes the medium in which it is flowing, determine the lift and drag that they will generate.

Figure 3.12 illustrates a basic airfoil shape and the forces that it generates. In this case, there is a thrust applied to a propeller. The lift is generated perpendicular to the angle of attack. Lift occurs because of the fact that the air traveling over the top of the airfoil is traveling a longer distance than the air along the bottom of the airfoil. Since the time for Savonius rotor's air to travel is the same, the air on the top of the airfoil is traveling at a higher velocity, and therefore there is a lower pressure on the top; thus, lift occurs. Drag is the frictional force exerted on the airfoil.

The flow of air as just discussed spins the propeller with an efficiency that can ultimately approach what is referred to as the "Betz's constant." This constant is derived from the kinetic energy formula as applied to the kinetic energy stored in the area of the wind stream where the wind turbine is located, and the application of Bernoulli's equation. For example, kinetic energy (KE)

$$\text{Kinetic energy} = (1/2) \times \text{Mass} \times \left(\text{Velocity}^2\right) \tag{3.7}$$

This equation can then be altered to suit the flow of air in the following manner:

$$\text{KE} = (1/2) \times \text{Air density} \times \text{Area} \times \left(\text{Velocity}^2\right) \tag{3.8}$$

In addition, since Power = KE ÷ Time, the equation can be further refined to:

$$\text{Power} = (1/2) \times \text{Air density} \times (\text{Distance} \div \text{Time}) \times \text{Area} \times \left(\text{Velocity}^2\right) \tag{3.9}$$

Since distance divided by time is equal to velocity, then the equation becomes;

$$P = 0.5 \, \rho A V_w^{3q} \tag{3.10}$$

$$\text{Power} = P = (1/2) \times \text{Air density} \times \text{Area} \times \left(\text{Velocity}^3\right)$$

where,

P = available power in the wind
ρ = air density = 1.2 kg/m^3
A = the area of the wind turbine blades (m^2)
V_w = velocity = wind speed (m/s)

3.4 WIND POWER PROFILE

The power available from the wind varies as the cube of the wind speed. If the wind speed doubles, the power of the wind (the ability to do work) increases 8 times. For example, a 10-mile per hour wind has one eighth the power of a 20-mile per hour wind: $10 \times 10 \times 10 = 1000$ versus $20 \times 20 \times 20 = 8000$.

One of the effects of the cube rule is that a site, which has an average wind speed reflecting wide swings from very low to very high velocity, may have twice or more of the energy potential of a site with the same average wind speed that experiences little variation. This is because the occasional high wind packs a lot of power into a short period of time. Of course, it is important that this occasional high wind come often enough to keep batteries charged.

If you are trying to provide smaller amounts of power consistently, you should use a generator that operates effectively at slower wind velocities. Wind speed data are often available from local weather stations or airports, as well as the US Department of Commerce, National Climatic Center in Asheville, NC. You can also do your own site analysis with an anemometer or totalizer and careful observation. Installation of generators should be close to the battery bank to minimize line loss, and 20 feet higher than obstructions within 500 feet. The tower should be well grounded.

In general the mechanical power output of a wind turbine, P_m, is given as:

$$P_m = \left(\frac{1}{2}\right) \times \rho \times \left(V^3\right) \times A \times C_p \tag{3.11}$$

where C_p = the aerodynamic power coefficient, which represents the efficiency of the wind turbine.

The maximum value of C_p is $16/27 = 0.593$. This maximum value of $16/27$ was first pointed to by Betz and is commonly recognized as the ceiling for wind turbine performance.

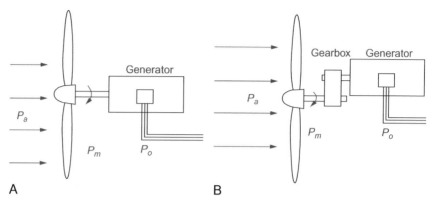

A   B

FIGURE 3.13 Variation of power coefficient with blade tip speed to wind speed ratio.

$$P_{\text{max}} = 0.593 \times P_a \qquad\qquad (3.12)$$

P_{max} = the maximum extractable power from the wind through the wind turbine, or it is the maximum mechnical power available to the generator
P_a = the available power in the wind

This means that no wind turbine could convert more than ~59% of the kinetic energy available in the wind. Realizing this makes the modern three-bladed, horizontal axis machines' capability of achieving approximately 45% efficiency all the more impressive.

C_p, the aerodynamic power coefficient, depends on the type of the windmill as it is shown in Figure 3.13. The power coefficient is a function of the blade tip speed to wind speed ratio λ. This is also called the extractable generator power from the wind. The instantaneous power output of the generator is given as P_o:

$$P_{\mathbf{o}} = \left(\frac{1}{2}\right) \times \rho \times \left(V^3\right) \times A \times C_p \times \eta \qquad\qquad (3.13)$$

where η is the efficiency of the generator if no gearbox is used as in Figure 3.14(a) and the combined gearbox generator efficviency if there is a gearbox as in Figure 3.14(b). However, the average value of the output power of the generator can be written as:

$$P_{\text{av}} = \int_0^\infty P_o f(v)\,dv \qquad\qquad (3.14)$$

where $f(v)$ is the probability density function.

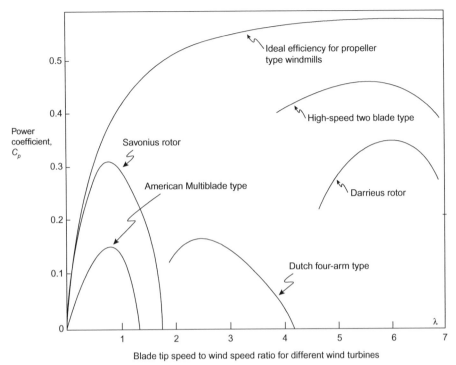

Blade tip speed to wind speed ratio for different wind turbines

FIGURE 3.14 The available, pechanical, and output power of a WECS: (a) Wind turbine gearbox generator and (b) wind turbine generator.

EXAMPLE 3.4

A wind energy system has the following parameters:

- Rotor turbine diameter, $D = 37.5$ m
- Wind turbine maximum power coefficient, $C_{po} = 0.38$
- Cut-in wind speed, $V_{wi} = 5.4$ m/sec
- Rated wind speed, $V_{wr} = 9.7$ m/sec
- Cut-off wind speed, $V_{wo} = 17.9$ m/sec
- Air density, $\rho = 1.225$ kg/m^3
- Synchronous generator rated power output, $P_o = 200$ KW

a) Calculate the available power in the wind at rated wind speed:

$$P_a = 0.5\rho A V_{wr}^3 = 0.5\rho \Pi (D/2)^2 V_{wr}^3$$
$$P_a = 0.5 \times 1.225 \times 3.14 \times 18.75^2 \times 9.7^3 = 617.4 \text{ KW}$$

(Continued)

EXAMPLE 3.4—cont'd

b) If the wind turbine operates at maximum power coefficient at rated wind speed, then calculate the mechanical power extracted from the wind at rated wind speed, P_m:

$$P_m = P_a \times C_{po} = 617.4 \times 0.38 = 234.61 \text{ KW}$$

c) Calculate the combined Gear generator efficiency, η

$$\eta = P_o/P_m = 200 \div 234.61 = 85.25\%$$

d) Calculate the generator capacity factor, CF, if the average power output of the generator is $P_{av} = 120$ KW

$$CF = P_{av}/P_o = 120 \div 200 = 0.6$$

Usually the generator does not start producing electricity until the available power in the wind exceeds the power needed to overcome the windage and friction losses in the whole system. The wind speed at which the generator starts producing electricity is called *cut-in wind speed*, $V_{cut\text{-}in}$ (see Figure 3.15).

As the wind turbine speed increases, the power output of the generator increases until the generator produces its rated power output. The wind speed at which the generator produces its rated output is called *rated wind speed*, V_{rated}.

When the wind speed reaches a point where the wind turbine is rotating at a dangerously high speed, with the possibility of destruction due to the centrifugal forces,

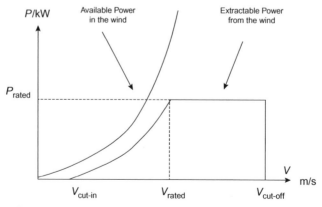

FIGURE 3.15 Wind speed power output profile.

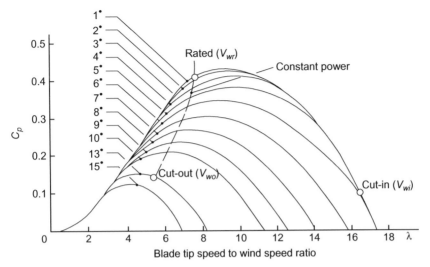

FIGURE 3.16 Mode-1 performance curves.

then the wind turbine is furled and the generator is stopped. The wind speed at which the generator is stopped is called cut of wind speed, $V_{cut-off}$.

For a synchronous generator to spell out the extra power after the generator reaches it rated output power pitch angle, control is used. The total blade is turned around using hydraulic power this is called *full pitching* or the tip of the blade is rotated and then this is called partial pitching. By rotating the blades the pitch angle is controlled and thus the power coefficient is controlled. Each pitch angle is represented by one curve, as seen in Figure 3.16. The power coefficient of wind turbines with a fixed-pitch angle is presented with only one curve.

3.5 BASIC ELEMENTS OF THE WECS

The wind energy conversion system (WECS) consists of the following parts:

- Wind turbine
- Wind generator
- Control system
- Gearbox
- Nacelle
- Instrumentation
- Support tower

A brief description of components is followed by detailed information about them. A cross-section of a Mode-1 is shown in Figure 3.17; in addition, Figure 3.18 shows the various components of a WECS in detail.

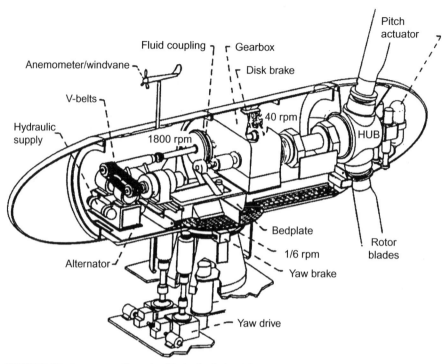

FIGURE 3.17 A cross-section of a Mode-1 wind turbine.

Wind turbine: This works the opposite of a fan. Instead of using electricity to make wind, as a fan does, wind turbines use wind to make electricity. The wind turns the blades that spin a shaft, which connects to a generator and makes electricity. It consists of a number of blades and a hub or rotor to which the blades are attached; the hub could be rigid or teetered, which see-saws as it rotates. This allows the stress in the rotor to be relieved. The blades of the rotor are constructed of fiberglass, wood, steel, aluminum, or titanium. The number of blades ranges from 1 to multiple, depending on the application. The turbine is designed to capture the wind's energy and covert it into mechanical energy to drive the generator to produce electricity. Wind turbines are classified into vertical axis and horizontal axis; they are also classified as downwind and upwind.

Generator: This is connected to the wind turbine directly through a shaft or through gearing. When the input speed is about the cut in, energy is produced. The generator is optimized to produce its rated output power at the rated wind speed. It produces electricity until the wind speed climbs to cut-off speed, where the generator is stopped by braking and the wind turbines are feathered to prevent overspeed damage to it and the turbine. The generator consists of

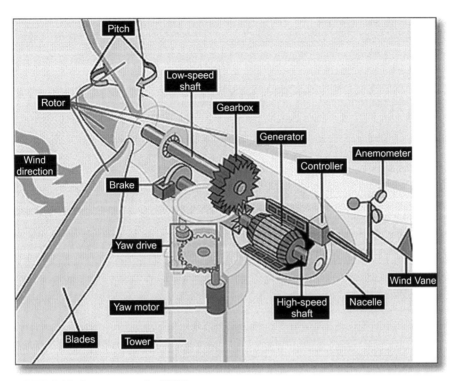

FIGURE 3.18 Components of a WECS.

both a rotating rotor and a stationary stator. Common generator types include the DC, AC induction, and AC synchronous.

DC: The advantage of this generator is that it does not require a rectifier for DC voltage. The disadvantage is that it is very heavy, thus requires a stronger support tower. Another big disadvantage is the extra maintenance involved in replacing the brushes every six months.

AC induction: This generator uses the very rugged squirrel cage rotor. Pitch angle or braking must be used to keep the rotor from spinning above maximum torque, thus destroying the generator. A major advantage of this generator over the synchronous one is that no synchronization is required. A disadvantage is that it needs sources of reactive power to self-excite.

AC synchronous: A disadvantage of this generator is that excess wind turbine speed has to be spilled out to keep the electromagnetic speed synchronized to the rotor speed. This is usually done with expensive pitch angle control. A major advantage is that this generator produces reactive power from which so many appliances operate. Recently brushless synchronous generators have been used. They run at variable speeds and do not need pitch–angle control.

Control systems: WECS are designed for fail-safe unmanned operation. The following are several control options:

Yaw control: Horizontal axis machines require yaw control; that is, a mechanism is used to swing the turbines into line with the wind. The yaw mechanism consists of an electric motor and a gearbox to point the turbine in the direction of the wind. Small machines usually have a tail assembly to do that. Large machines usually have a servo mechanism that orients them to the direction of maximum power.

Pitch–angle control: This is used to maintain a certain shaft speed that is constant by increasing or decreasing turbine thrust. For this, both full and partial pitching is used. Partial pitch–angle control was used in Mode-2 of the federal wind energy turbines series (Figure 3-19). A pitch actuator is used to control hydraulic flow and thus rotate the blades. In full pitch–angle control, the whole blade is rotated; in partial pitching. only the tip of the blade is rotated.

Braking: It is used for yaw and turbine control. The brake is activated for maintenance or for emergency situations.

Computer/microprocessor control: This is also used to determine proper shaft speed for optimum output power. The WECS runs as unmanned—the microprocessor starts and stops and control the system.

Modern wind turbines are usually equipped with mechanisms to prevent damage in excessively high winds. Large machines may have complex arrangements to shut down generation at high wind speeds. Smaller systems change the

FIGURE 3.19 Partial pitch–angle control in a Mode-2 wind turbine.

blade orientation so that they present a smaller surface to the wind and thereby reduce the speed of rotation, or they use mechanical brakes.

Gearbox: It is used to maintain a fixed or variable ratio between the shaft and the generator speed. The gearbox, or transmission, steps up the speed of rotation of the wind turbine to a suitable speed required by the generator of up to 1800 rpm. With a gearbox you can convert a slowly rotating, high-torque power from the wind turbine rotor to a high-speed, low-torque generator.

Nacelle: This houses the generator and instrumentation. Usually, it is made of fiberglass and located on the top of a tower.

Instrumentation: An anemometer/wind vane is used to measure wind speed for an input to the microprocessor. It sends a signal to the microprocessor to allow the turbine to turn. In addition, lightning protection and arrestors protect the generator. Various meters, such as kW meter, kWh meter; ampere meter, and voltmeter, are installed in a WECS.

Support tower: The tower places the wind turbine well above the turbulent air currents that are close to the ground and captures the higher wind speed. Generally, the turbines should be 50 feet above the nearest obstacle. Tower design is particularly critical because it must be robust enough to sustain the compression and the drag forces of the WECS. Also the tower should be designed to allow access for maintenance. A particularly important aspect of tower design is elimination of resonance between the frequency range of the rotating blades and the resonant frequency of the tower. Gye wires are normally used to help support the tower and reduce forces. Towers are generally made up of galvanized steel. Three popular tower structures are used.

Tubular steel. Most large wind turbines are delivered with tubular steel towers.

Lattice. The advantage of lattice towers is their cost effectiveness. A disadvantage is their visual appearance.

Guyed pole: Many small wind turbines are built with narrow pole towers supported by guy wires.

Choosing between low and tall towers: Generally it is an advantage to have a tall tower in areas with high terrain roughness. Short towers increase the fatigue loads on the turbine.

Manufacturers often deliver machines where the tower height is equal to the rotor diameter.

3.5.1 **Types of wind turbines**

During the 200 years of its development has led wind turbines to be manufactured and built in many types. Wind turbines differ in their axis of rotation, differ in whether they are upwind or downwind, differ in their sizes and shapes, differ in whether the blades have a fixed pitch–angle or a variable one.

Of the many types of wind turbines, the fixed pitch are the simplest. This is where the blades have a fixed angle built into the turbine and are generally a one-piece construction. A less common ground-adjustable wind turbine operates on the same principle as the fixed pitch but is able to be adjusted. This is done by loosening the clamping mechanism, which holds the blades. The blades can be adjusted using a protractor to measure angles. A controllable-pitch wind turbine allows a pitch–angle change while the turbine is rotating. The turbine can be adjusted anywhere between its minimum and maximum allowable angles. A governor normally is used to maintain a certain speed by adjusting the blade through hydraulic fluid flow. Hydraulic pressure is used to increase pitch and counterweights are used so that centrifugal force decreases the angle.

Small wind turbines are used for providing power off the grid, ranging from very small, 250-W turbines designed for charging batteries on a sailboat, to 50-kW turbines that power dairy farms and remote villages. They have tail fans that keep them oriented into the wind. Large wind turbines, used by utilities to provide power to a grid, range from 250 kW up to the enormous multi-MW machines that are being tested in Europe. Large turbines sit on towers that are up to 60 m tall and have blades that range from 30 to 50 m long.

Utility-scale turbines are usually placed in groups or rows to take advantage of the prime windy spots. Wind farms can consist of a few or hundreds of turbines to provide enough power for whole towns. Electricity must be produced at just the right frequency and voltage to be compatible with a utility grid. Since wind speed varies, the speed of the generator could vary, producing fluctuations in the electricity. One solution to this problem is to have constant speed turbines, where the blades adjust, by turning slightly to the side, to slow them down when wind speeds gust. Another solution is to use variable-speed turbines, where the blades and generator change speeds with the wind, and sophisticated power controls fix the fluctuations of the electrical output. A third approach, adopted by only one company so far, is to use low-speed generators. An advantage that variable-speed turbines have over constant-speed turbines is that they can operate in a wider range of wind speeds.

Modern electric wind turbines come in a few different styles, depending on their use. The most common styles are described in the following subsections.

Horizontal axis wind turbine (propeller type)

The horizontal axis wind turbine (HAWT) has the axis of the blades horizontal to the ground. On this turbine, two or three, or multiple, blades spin upwind of the tower that it sits on. Great care is taken about the design of the rotor tip because the tip of the blade moves substantially faster than the root of the blade. Blade tips have changed over time with continuing research, which is also done to study performance, since most of the torque of the rotor comes from the outer part of the blades. In addition, the airflow around the tip of rotor blades is extremely complex, compared to the airflow over the rest of the rotor blade.

Modern wind turbine engineers avoid building large machines with an even number of rotor blades. The most important reason is the stability of the turbine. A rotor with an odd number of blades can be considered to be similar to a disc when calculating the dynamic properties of it. A rotor with an even number of blades causes stability problems for a machine with a stiff structure. The reason is that at the very moment when the uppermost blade bends backward, because it gets the maximum power from the wind, the lowermost blade passes into the wind shade in front of the tower.

Two-bladed: This wind turbine design has the advantage of saving the cost of one rotor blade and its weight. However, they tend to have difficulty in penetrating the market, partly because they require a higher rotational speed to yield the same energy output. In addition, they have a teetering effect as they rotate; this is a disadvantage both in regard to noise and visual intrusion.
One-bladed: This type of wind turbines does exist and saves the cost of another rotor blade. One-bladed wind turbines are not widespread. They have the same problems, as mentioned under the two-bladed design, and apply to an even larger extent. In addition to higher rotational speed and the noise and visual intrusion problems, they require a counterweight to be placed on the opposite side of the hub from the rotor blade in order to balance the rotor. So there is no weight-savings compared to a two-bladed design.

Most modern wind turbines use three-bladed designs with the rotor position maintained upwind by using electrical motors in their yaw mechanism. The majority of the turbines manufactured today have this design. Widely used HAWTs come in different configurations, as follows:

- Three-blade turbines are shown in Figure 3.20. Most grid-connected commercial wind turbines today are built with a propeller-type rotor with three blades. Wind flows over a blade that is at an angle to the direction of the wind flow. The blade is a propeller that uses existing aerodynamic theory as applied to aircraft. The angle in which it is positioned to the wind stream is referred to as the *angle of attack*.

FIGURE 3.20 Example of a three-bladed horizontal axis wind turbine.

FIGURE 3.21 Picture of a Dutch wind turbine.

- American multiblade wind turbines are shown in Figure 3.2. It usually is a small type, more suitable for water pumping than generating electricity because it has a high torque due to its high solidity as a result of the high number of blades.
- The Dutch wind turbine is shown in Figure 3.21. It consists of two crossed arms. Again it is more suited for water pumping than generation of electricity.

Vertical axis wind turbine

A second style of wind turbine is the vertical axis wind turbine (VAWT) design. The axis of the blades is vertical to the wind stream.

- The Darrieus type is shown in Figure 3.22. It is the only vertical axis turbine that has ever been manufactured commercially at any volume. Named after the French engineer Georges Darrieus who patented the design in 1931, it is characterized by its C-shaped rotor blades, which make it look a bit like an eggbeater. It is normally built with two or three blades. His style of vertical axis machine uses aerodynamically shaped wings to add some lift effect that is generated from the airfoil blades during rotation. In general, the Darrieus machines will achieve a slightly higher tip-speed to wind-speed ratio than a horizontal axis machine, but their power coefficient is less than a comparable horizontal-axis machine.
    - *Advantages:* The gearbox is located on the ground, it is easy to maintain, a tower isn't needed to support the system, and there is no need for a yaw mechanism to turn the rotor against the wind.

FIGURE 3.22 Darrieus-type vertical-axis wind turbine.

- • *Disadvantages:* Low efficiency because the machine is not self-starting, may need impractical guy wires to hold it up, and replacing the main bearing for the rotor is difficult.
- The empirically measured power coefficient of commercially built Darrieus machines reach up to approximately 35%, even with mechanical and frictional losses present. A Darrieus machine does not generate the low-speed torque that is typical of other vertical-axis machines. This does introduce the problem of the machine not being able to start itself; it requires an external push to get it rotating. The design also poses other aerodynamic issues; that is, the downwind rotor is always in the turbulent wake of the upwind rotor and the tower, the torsion forces on the blades are very asymmetric and erratic, and the rotors have uneven loading from top to bottom because of the proximity of ground turbulence.
- They also suffer from a variety of problems that have limited their use. The first is that the power of the wind increases with height from the ground and Darrieus blades are low to the ground. In addition, the blades have been made with aluminum, which weakens with stress. Because the blades are weak, they cannot capture the higher-powered, high-speed winds. Some of these shortcomings could be minimized with further development and as better materials are invented.
- The Savonius-type wind turbine is shown earlier in Figure 3.11.
- Other types of vertical-axis wind turbines are shown in Figures 3.23 through Figures 3.29.

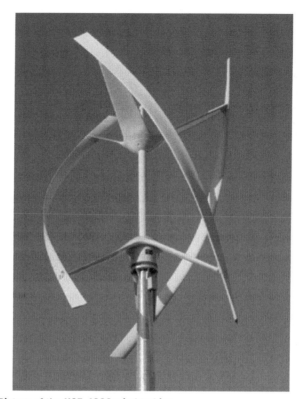

FIGURE 3.23 Picture of the UGE-4000 wind turbine.

FIGURE 3.24 The S-Sail type of wind turbine.

FIGURE 3.25 Windspire vertical-axis wind turbines.

FIGURE 3.26 Illustration of a cross-flow wind turbine.

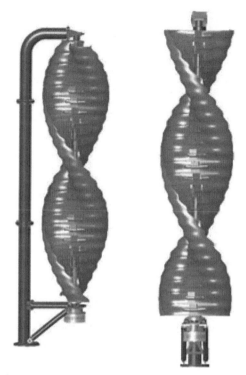

FIGURE 3.27 Illustrations of helical-type wind turbines.

3.5.2 **HAWT versus VAWT**

The *HAWT* has the following advantages:

- Can be started easily
- Higher efficiency than VATB
- Easy to control

The following are some of its disadvantages:

- Need strong tower to support the generator and the nacelle
- TV interference
- Difficult to maintain

The *VAWT* has the following advantages:

- Generator gearbox on the ground is easy to maintain
- Less TV interference
- Do not need a tower
- Do not need a yaw mechanism to turn the rotor against the wind

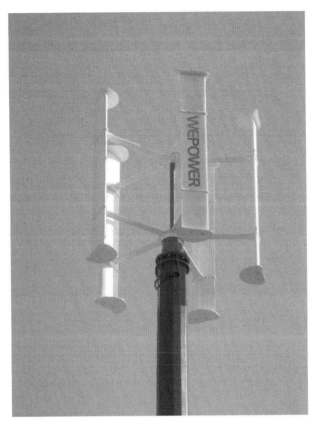

FIGURE 3.28 The WePower Falcon series type wind turbine.

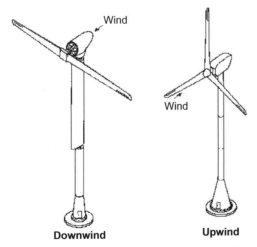

FIGURE 3.29 Illustrations of downwind- and upwind-type of wind turbines.

The following are some of the VAWT's disadvantages:

- Less efficient
- Difficult to start and control
- Needs guy wire
- Blades subjected to uneven loads that cause a lot of stress to them

The installation of the wind turbine with respect to the tower takes two shapes, whether the wind hits the turbine first or the tower first. There are two different arrangements: upwind and downwind:

3.5.3 Upwind and downwind machines

Upwind machines have the rotor facing the wind; this is the more popular design. The wind hits the tower first (Figure 3.30). The basic advantage of upwind designs is that one avoids the wind shade behind the tower. In addition, an upwind machine needs a yaw mechanism to keep the rotor facing the wind.

Downwind machines have the wind hitting the turbine first (see Figure 3.29). They have the theoretical advantage that they may be built without a yaw mechanism if the rotor and nacelle have a suitable design that makes the nacelle follow the wind passively; this is doubtful for large wind turbines.

3.5.4 Large turbines versus small turbines

The following are some reasons for choosing large turbines:

- Are usually able to deliver electricity at a lower cost than smaller ones
- Are well suited for offshore wind power
- A large turbine, with a tall tower, uses the existing wind resource more efficiently

Some of the big turbines on the market follow, with brief descriptions:

- *NEG Micon 2 MW:* It has a 72-m diameter and is mounted on a 68-m tower
- *Bonus 2 MW:* It has a 72-m rotor diameter and is mounted on a 60-m tower
- *Nordex 2.5 MW:* Its rotor diameter is 80 m and it is mounted on an 80-m tower; turbine has pitch power control

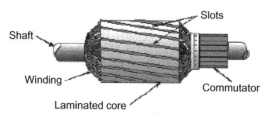

FIGURE 3.30 The rotor of a DC generator.

Some reasons for choosing smaller turbines are as follows:

- Local electrical grid may be too weak
- Less fluctuation in the electricity output than from a wind park
- Spreads the risk
- Aesthetical and landscape considerations

3.5.5 Wind turbine generators

The wind turbine generator is a device that converts mechanical energy into electrical energy. The typical operation of the wind turbine generator is that it has to work with a power source that supplies fluctuating mechanical power. There are a few types of generators that can be used as turbine generators (e.g., DC, induction, synchronous, and switch reluctance). Large manufacturers can supply both 50-Hz and 60-Hz wind turbine generator models.

Principles of DC generators

The DC generator consists of the following parts:

Stator: This consists of magnetic poles around which the field winding is rapped. Is is a term that is commonly used to describe the stationary component of a DC machine. The field winding provides the magnetic field with which the mechanically rotating (armature or rotor) member interacts.

Armature (rotor): Made up of a stack of laminations, it is the rotating component of a DC generator. The armature has slots imbedded in the armature winding, which is made of copper wire turns. The winding is made up of many coils and the ends of each coil are connected to a copper segment. The armature or rotor is the only rotating part. It is made of thin, highly permeable and electrically laminated steel stacked together and rigidly mounted on the shaft.

Commutator: A cylindrical device mounted on the armature shaft, it consists of a number of wedge-shaped copper segments arranged around the shaft and is insulated from it and each other. The armature coils are terminated with these segments.

Brushes: The motor brushes ride on the periphery of the commutator and electrically connect and switch the armature coils to the power source. The brushes are held in a fixed position against the commutator at a certain pressure. The brush is a piece of current-conducting material (usually carbon or graphite) that rides directly on the commutator and conducts current from armature windings to the load; that is, it is the electrical connection between the armature coils and the external circuit.

Figure 3.30 shows the rotor of a DC generator. Figure 3.31 shows a cross-section of a DC machine.

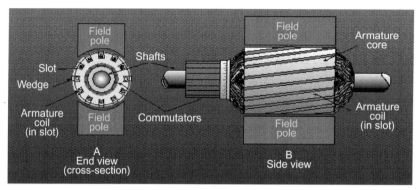

FIGURE 3.31 A cross-section of a DC machine.

Operation

The field current produces a magnetic field and the wind turbine rotates the rotor (armature) of the DC generator. The magnetic field flux lines cut the armature winding coils and by Faraday's Law generate in them an AC current and voltage, which are rectified by the commutator into DC current and voltage and, through the brushes, the current flows to the load.

The field windings are wound onto the poles so that they in polarity. How the field windings are wound determines the type of DC machine. If the field winding is connected in series with the armature winding, then it is a series machine. If the field winding is in parallel with the armature winding, then it is a shunt machine. However, usually the shunt generator is used in wind energy conversion systems. When the DC shunt machine is used as a generator, the starting field flux is built from residual magnetism. Sometimes the field winding is replaced by a permanent magnet, then the generator is called DC permanent magnet generator; this is what is used for small wind turbine generators.

The wind turbine rotates the rotor of the DC generator. The driving torque, T_d, developed by a generator is equal to the machine's constant, K_a, times the magnetic flux, Φ, times the armature current, I_a. Assuming magnetic linearity (i.e., the field is proportional to the current passing thorough the field winding), Eq. (3.15) applies:

$$T_d = K_a \Phi I_a \tag{3.15}$$

The open-circuit armature voltage produced across the brushes, E_a, is given as

$$E_a = K_a \Phi \omega_m \tag{3.16}$$

The power output of the DC generator, P_o, is calculated as

$$P_o = V_t I_L \tag{3.17}$$

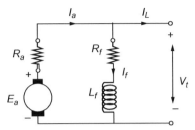

FIGURE 3.32 The equivalent circuit of the shunt DC generator.

where I_L is the line current and V_t is the terminal voltage that is calculated as

$$V_t = E_a - I_a R_a \tag{3.18}$$

The equivalent circuit of a shunt DC generator is shown in Figure 3.32.

The DC generator can find its application in the area where the load requires only direct current. If an AC load is required, then an inverter must be introduced between the generator and the load. Even though the DC generator is suitable for delivering DC current, its complex construction may yield high cost, and it is a heavy machine that is not well suited for wind turbine applications. Moreover, the brushes need to be changed periodically, which increases it maintenance cost. Also, the commutator is made of copper, which is an expensive material. Furthermore, the DC generator does not have the ruggedness and service ability. Therefore, it can be difficult to warrant its application in wind generation.

Principles of AC generators

A set of three phase–current systems fed to a three-phase wind system, the stator of an AC machine will always produce a rotating magnetic field (Figure 3.33). The speed of the rotating magnetic field is called synchronous speed, ω_s, measured in rad/sec, or n_s, measured in revolutions per minute, where

$$n_s = 120 f/p \tag{3.19}$$

f is the frequency and p is the number of poles. Also,

$$\omega_s = (n_s \div 60) \times 2\Pi \tag{3.20}$$

The AC machine consists of a stationary stator with a set of winding and rotating rotors with another set of winding. The speed of the rotor is ω_m in rad/sec, or n_m, in revolutions per minute. If $\omega_s = \omega_m$ and $n_s = n_m$, then the AC machine is called synchronous; if they are not equal, then the AC machine is called asynchronous.

Because of the difference in the rotor speed and the synchronous speed in asynchronous machines the rotating magnetic field, produced by the stator windings induces voltages and currents in the rotor windings. For this reason the asynchronous machine is also called induction machine.

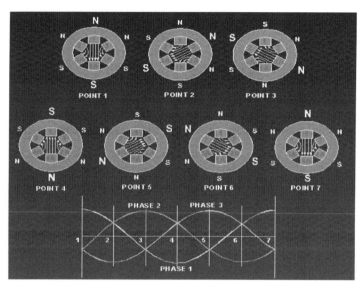

FIGURE 3.33 The rotating magnetic field in AC machines.

Synchronous generators

The synchronous generator is a type of AC machine commonly used for wind power generation. It runs at a speed that precisely corresponds to the frequency of the supply. Furthermore, the frequency of the voltage and current in the generator correspond exactly to the speed of the synchronous speed. If the synchronous generator is connected to a supply network, its rotational speed has to be exactly constant. Synchronous generators consist of a stationary stator with slots imbedded in the stator windings, which is also called the armature windings and/or a rotating rotor. There are two versions of the rotor's design.

In the first version, the rotating rotor is equipped with a DC current-carrying winding, the DC current produces the magnetic field, and the rotor winding is also called the excitation winding or the field winding. The winding is imbedded in the rotor slots. A generator with such a design is called *regular synchronous* (Figure 3.34). The rotor has no winding in the second version; the winding is replaced by a permanent magnet that produces the magnetic field. This type of generator is called *permanent magnet synchronous.*

Regular synchronous generators. Regular synchronous generators (RSGs) use a DC current fed to the rotor winding to generate the magnetic field. The stator has three phases of winding a, b, and c imbedded in its slots, and it has the rotor winding that is also called the *field winding* or excitation winding. When the rotor is driven by the wind turbine, the magnetic field-flux lines cut the stator windings and induce in them AC voltage called the armature voltage, E_a.

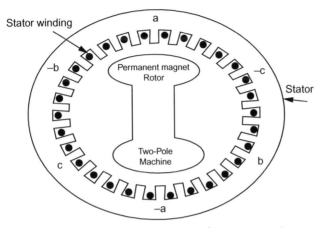

FIGURE 3.34 Cross section of a permanent magnet synchronous generator

This voltage is proportional to the magnetic field flux, Φ, the stator frequency, f, and the number of turns of the stator winding, N.

$$E_a = k\,\Phi\,N\,f \qquad (3.21)$$

The equivalent circuit of a synchronous generator per phase is shown in Figure 3.35. The terminal voltage, V_a, is calculated as

$$V_a = E_a - I_a/\Theta(R_a + jX_s) \qquad (3.22)$$

where,

R_a = the ohmic resistance of one phase
X_s = the synchronous reactance
Θ = the power factor angle, which depends on the load

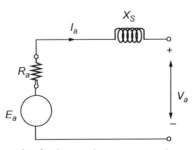

FIGURE 3.35 The equivalent circuit of a synchronous generator.

The active power output, P_o, is given as

$$P_o = 3V_a I_a \cos\Theta \tag{3.23}$$

The reactive power generated, Q_o, is given as

$$Q_o = 3V_a I_a \sin\Theta \tag{3.24}$$

However, regular synchronous generators need DC excitation current, thus need an exciter that produces the DC current. In addition, they are difficult to synchronize with the existing power system. To operate synchronous generators, the rotor speed of them must be kept constant and the pitched-angle control must be used to control the speed of the turbine.

Permanent magnet synchronous generators. Permanent magnet synchronous generators (PMSGs) replace the DC winding on the rotor with a permanent magnet. This magnet is a powerful one, made from rare earth metals (e.g. Neodynium), and by virtue of being a magnet, it produces a constant magnetic flux. The benefit of the PMSG over the regular one is that there is no need for a DC current source and thus no need for an exciter.

The disadvantage of the PM synchronous generator is that the rare earth magnets are very expensive, therefore the inclusion of the word "rare" in the name. The overall disadvantage of the synchronous generator is that the frequency of the generated voltage is directly proportional to the rotational speed of the prime mover. This is not significant if you are using a synchronous generator with a steam turbine that can be very precisely controlled. The problem is the random nature of the wind's speed, and that there are currently no methods of controlling the wind turbine that can react as fast as the wind can change.

Figure 3.36 shows a cross-section of the brushless synchronous generator. It consists of the following:

Stationary stator: This is the stationary component; it is made up of a stack of lamination and armature coils similar to that of regular synchronous generator.
Rotor: This is the rotating component of brushless motor that is made up of rare earth magnets, which provide the magnetic field.
Encoder resolver or hall sensor: The encoder is a feedback device used for commutation purposes.

In general, synchronous generators can be either nonpermanent (regular) or permanent magnetic types and both generate reactive power. The permanent magnetic generators are brushless generators and have similar characteristics to DC generators; this is why they are also called *brushless DC generators*.

Induction generators

An induction generator is an alternating current generator in which the primary winding on one member (usually the stator) is connected to the power system and a secondary winding on the other member (usually the rotor) carries the

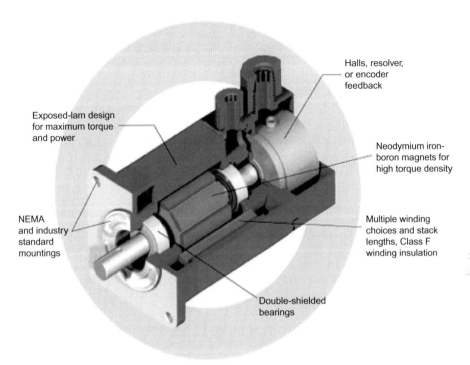

Halls, resolver, or encoder feedback

Exposed-lam design for maximum torque and power

Neodymium iron-boron magnets for high torque density

NEMA and industry standard mountings

Multiple winding choices and stack lengths, Class F winding insulation

Double-shielded bearings

FIGURE 3.36 A cross-section of a permanent magnet synchronous generator.

FIGURE 3.37 The rotor of a squirrel-cage induction machine.

induced current. The rotor could be a squirrel-cage type (Figure 3.37) or the rotor carries windings that are a replica of the stator windings. These winding are terminated with slip rings for speed change. This kind of induction machine's design is called a *wound rotor machine*. Most wind generators in the world use induction generators to generate alternative current. This type of generator is not widely used outside of the wind turbine industry. The major difference between the synchronous generator and induction generator is that the induction one cannot generate reactive power.

Principles of operation and components

In an induction motor, the set of the three phases of current flowing in the stator produces a rotating magnetic field; the flux lines of the rotating magnetic field cut the rotor windings and induce in them currents and voltages. The set of the three induced currents in the rotor produces a rotating field. The two magnetic fields interact and the motor runs (Figure 3.38). If the rotor of an induction machine is spun by the wind turbine at a speed higher than the synchronous speed, $\omega_m > \omega_s$, the motor becomes a generator. The following list contains construction information.

Stator: A stationary portion of the induction generator made up of a stack of lamination and an armature
Winding: The set of the three phase windings is connected to the power system. The stator of an induction generator is similar to the stator of a synchronous generator.
Rotor: The rotating portion of an induction generator is made up of a stack of lamination. It is either wound type or squirrel-cage type. The equivalent circuit of the induction generator is shown in Figure 3.39; definitions follow:
- R_1 and R_2 are ohmic resistances of the stator and rotor, respectively, referred to the stator
- X_1 and X_2 are the stator and rotor leakage reactances, respectively, referred to the stator
- R_c is the core loss resistance referred to the stator
- X_m is the magnetizing reactance referred to the stator
- I_1 and I_2 are the stator and rotor current, respectively, referred to the stator
- V_1 is the stator phase terminal voltage
- S is the slip

FIGURE 3.38 A cross-section of the induction generator.

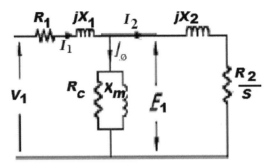

FIGURE 3.39 Equivalent circuit of the induction generator.

Because the synchronous speed and the rotor speed are not equal to the difference between them in percentage points, this is called the slip. It is expressed as a percentage of the synchronous speed.

$$S = (n_s - n_m)/n_s \tag{3.25}$$

The power output, P_o, is given as

$$P_o = 3V_1I_1\text{Cos}\Theta \tag{3.26}$$

The power that crosses the air gap from the rotor to the stator or vice versa is called air gap power, P_g:

$$P_g = 3I^2{}_2R_2/s \tag{3.27}$$

The developed power, P_d, is:

$$P_d = (1 - s)P_g \tag{3.28}$$

The developed torque, T_d, is given as

$$T_d = P_d \div \omega_m = P_g/\omega_s \tag{3.29}$$

The torque speed characteristics of an induction machine are shown in Figure 3.40. For the induction generator, the rotor speed should be more than the synchronous speed, $\omega_m > \omega_s$, and the slip is, of course, negative.

The rotor should move faster than the rotating magnetic field and the stator induces a strong current in the rotor. Furthermore, the speed of the induction generator will vary with the turning force applied to it. This is called the *generator's slip*. The speed of the generator will be increased or decreased slightly if the torque varies, which in turn causes less wear and tear on the gearbox. This is one of the most important reasons for using an induction generator rather than a synchronous generator on a wind turbine system.

Even though the induction generator is suitable for the system; it has some advantages and disadvantages. Induction generators need not to be synchronized

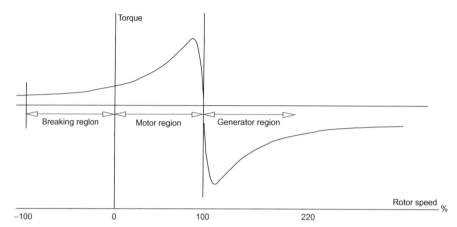

FIGURE 3.40 Speed torque characteristics of an induction machine.

and excited and they are inexpensive. They also do not need a stringent pitched-angle control. However, a source of reactive power must be used to operate the generator. The efficiency is good at full load, but falls off rapidly above and below that. The squirrel-cage rotor is normally a very rugged and low-maintenance option.

Another power-quality issue comes from the connection to the grid. When there is not sufficient wind speed to spin the generator past synchronous speed and produce power, the wind turbine remains disconnected from the power grid. If the wind were to pick up speed to the point where the generator could produce power, the generator becomes connected to the system. Another component of power quality that is a very important issue to utility companies is the power factor. Windmills are also equipped with switchable capacitor banks that are connected in shunt with the output of the turbine. These capacitors are controlled by the control system to compensate for lagging power factors.

The wound rotor induction machine could be used as an induction generator. In this case, a rectifier is connected to the rotor winding through the slip rings; it rectifies the rotor voltage and current. The rectified voltage is then connected to an inverter that inverts the DC rotor current and voltage to an AC voltage at the stator's frequency. The rectifier and inverter act like resistance, changing the rotor speed, but unlike a resistance, the energy from the rotator is recovered. This is called *slip power recovery*. It also is called double output induction generator (DOIG) because the generator has two outputs—one from the stator side and another from the rotor side.

In this configuration, the generator also generates more than its rated power output without being overheated because the stator and rotor currents are not allowed to exceed their rated values by changing the inverter firing angle. The configuration of the DOIG is shown in Figure 3.41.

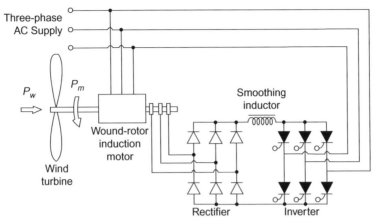

FIGURE 3.41 A schematic diagram of the DOIG.

EXAMPLE 3.5

A three-phase, 10 kVA squirrel-cage induction generator is driven by a fixed pitch–angle wind turbine. The generator delivers its rated output at a wind speed of 12 m/sec. The generator is connected to an infinite bus (i.e., a power system where frequency and voltage are held constant); the line-to-line voltage at the infinite bus is 208 V. The generator is driven at a rated wind speed, delivering its output at a lagging power factor of 0.8.

a) Calculate the active power output, P_o:

$$\Theta = 36.37, \cos\Theta = 0.8, \sin\Theta = 0.6$$
$$P_o = S \times \cos\Theta = 10 \times 0.8 = 8\,\text{kW}$$

b) Calculate the generator output current. The phase voltage is $V = 120$ V:

$$I = P/3\,V\,\cos\Theta = 8000 \div 3 \times 120 \times 0.8 = 27.78\,\text{A}$$

c) Calculate the reactive power the generator needs, Q:

$$Q = S \times \sin\Theta = 10 \times 6 = 6\,\text{kVAR}$$

d) The wind speed was reduced to 8 m/sec; the generator efficiency stayed the same but the aerodynamic coefficient became 80% of what it was. Calculate the power output of the generator:

$$P_{o1} = 8000 = 0.5\,A\rho V_{w1}^3 C_{P1} \times \varsigma$$
$$P_{o2} = X = 0.5\,AR\rho V_{w2}^3 C_{P2} \times \varsigma$$
$$P_{o1}/P_{o2} = V_{w1}^3 C_{P1}/V_{w2}^3 C_{P2}$$
$$P_{o2} = P_{o1} \times V_{w2}^3 C_{P2}/V_{w1}^3 C_{P1} = 8000 \times 8^3 \times 0.8 \times C_{P1}/12^3 \times C_{P1} = 1896.3\,\text{W}$$

3.6 WECS SCHEMES FOR GENERATING ELECTRICITY

There are many electrical schemes that can be adopted to generate electricity using the previously discussed generators. These schemes differ in output frequency and speed of the rotor. In general, there are four distinct electrical schemes, which are discussed in the following subsections.

3.6.1 Constant speed, constant frequency scheme

Constant speed, constant frequency (CSCF) scheme employs a regular synchronous generator with either full pitch–angle control, where the whole blades are rotated hydraulically to keep the rotor speed constant, or partial pitching, where the tips of the blades are rotated (Figure 3.42). This kind of scheme is not good because the hydraulic control is too slow, so the generators end up getting out of synchronism very frequently. One of the advantages of this scheme is that the synchronous generator generates reactive power and the regular synchronous generator is a well-known technology.

3.6.2 Almost constant speed, constant frequency scheme

The almost constant speed, constant frequency (ACSCF) scheme employs a squirrel-cage induction generator with less stringent pitch–angle control to keep the slip within 1 to 5%. (Figure 3.43). One of the advantages of this scheme is that the squirrel-cage induction generator is inexpensive and very rugged. A big disadvantage is that it needs reactive power to be self-excited; a capacitor across its terminal is used as a source of reactive power.

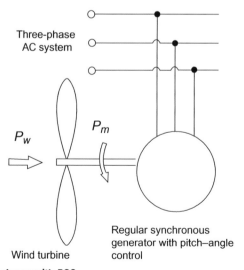

Three-phase
AC system

P_w

P_m

Wind turbine

Regular synchronous generator with pitch–angle control

FIGURE 3.42 CSCF scheme with RSG.

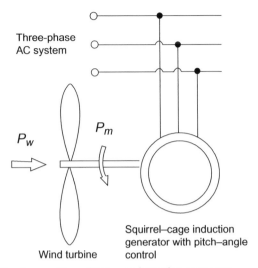

Three-phase
AC system

P_w

P_m

Squirrel–cage induction
generator with pitch–angle
control

Wind turbine

FIGURE 3.43 ACSCF scheme with squirrel-cage induction generator.

3.6.3 **Variable speed, constant frequency scheme**

Keeping the rotor speed of a WECS constant is a difficult task. Modern systems have turbines that run at variable speeds, eliminating the need for a costly and slow pitch–angle control. The variable speed, constant frequency (VSCF) scheme is also kept constant and the generator is still connected to the existing power system; to do that many schemes were used and some are described next.

The AC/DC/AC link scheme could employ a regular synchronous generator, a brushless synchronous generator, or the squirrel-cage induction generator (Figure 3.44). The wind turbines run at variable speeds; thus, the voltage and the frequency outputs of the generator are variables. To make them constant, a rectifier is used to rectify the voltage, then the DC voltage is inverted to a fixed frequency and voltage. This scheme is widely used; control is fast because it is done electronically not hydraulically.

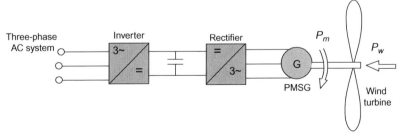

FIGURE 3.44 AC/DC/AC scheme with permanent magnet synchronous generator (PMSM).

By inverting the output of the wind turbine, the rotor speed is not an issue anymore, so a complicated control system is not needed. Another benefit of inverting the output voltage is that the DC power can be stored in batteries for times when there is not enough wind to supply the load. The batteries need to be connected to the DC link. The disadvantages of this type of scheme include the high-cost and reliability issues of the power electronics. Utility companies are not very fond of inverters because their output is not very clean due to harmonic distortion. To connect an inverter to a grid, it must have less than 5% total harmonic distortion.

The preceding is by far the most commonly used scheme. Other types of the VSCF schemes are:

- DOIG system (see Figure 3.40)
- Field moulated generator system
- AC-commutator generator system

3.6.4 Variable speed, variable frequency scheme

The variable speed, variable frequency (VSVF) scheme is used when the output frequency does not need to be constant; the generator in this case is working in a stand-alone mode and feeding frequency insensitive loads such as space and/or water heating. The system could employ a squirrel-cage induction generator, permanent magnet synchronous generator, or regular synchronous generator. There is no need for either pitch–angle or frequency control. The scheme is similar to that shown earlier in Figure 3.43.

3.7 SITING AND SIZING OF WECS

Wind energy is abundant. Wind resources are plentiful. Wind will not run out. Wind energy will be one of the most important, widely applied of the renewable energy forms during the next several decades. There are substantial challenges to be met, but all appear solvable. Successful research and development will potentially result in generation from wind energy of about 10% of the electricity used in the United States.

There are many wind characteristics that directly impact wind turbine performance. The uniqueness of the wind at each site indicates the need to measure speeds and profile local wind characteristics over multiple years. Examples of specific wind issues include:

- Wind energy increases dramatically with wind speed. The energy available in a 13 mph wind is more than double the energy available in a 10 mph wind.
- Wind speed generally increases with height above the ground. Therefore, modern, large wind turbines are more than 170-feet tall compared with old windmills, which were 20– to 30-feet tall.
- There are tremendous daily, seasonal, and yearly changes in the amount of wind energy available at a specific location.
- There can be large differences in available wind energy over a small area.

Besides the wind resource, wind producers have to deal with numerous issues related to: access to the site, environmental and avian concerns, ground conditions, and access to existing transmission lines. The higher costs associated with current renewable technologies represent another challenge facing organizations considering wind projects.

In industrialized countries, public acceptance of wind power is often the most important planning restriction and, consequently, also a political issue. Experience in developing countries is still limited, but recent large-scale applications in India and China show both reliable production and a high degree of public acceptance, along with private-sector participation.

Wind resources are best along coastlines and on hills, but usable wind resources can be found in most other areas as well. As a power source, wind energy is less predictable than solar energy, but it is also typically available for more hours in a given day. Wind resources are influenced by the ground surface and obstacles at altitudes up to 100 m. Thus, wind energy is much more siting-specific than solar energy. In hilly terrain, for example, two places are likely to have the exact same solar resource; however, it is quite possible that the wind resource can be different at both places because of site conditions and different exposure to the prevailing wind's direction.

Wind energy follows seasonal patterns that provide the best performance in the winter months and the lowest performance in the summer months; this is just the opposite of solar energy. To ensure economical operation, longevity, high efficiency, low maintenance, and seamless integration of wind power into the grid, a thorough understanding of the effect of all these issues at each site is critically important.

Sizing or matching windmills for a specific site is done based on the capacity factor. To correctly site a wind energy conversion system, it is helpful to consider the technical information and legal and environmental obstacles about an intended site's location, as detailed in the following.

SIZING OR MATCHING WINDMILLS TO SITES

Technical information
The following technical information is important when considering a site for the location of a WECS.

Average annual wind speed and prevailing direction
The wind's speed is the most important factor influencing the amount of energy a wind turbine can convert into electricity. Increasing wind velocity increases the amount of air mass passing the rotor; so, increasing wind speed will also have an effect on the power output of the system. The energy content of the wind varies with the cube (the third power) of the wind's average speed. Thus, if wind speed doubles, the kinetic power gained by the rotor increases eight times. In conclusion, a site should have a high average wind speed.

The site should have a prevailing wind; otherwise, a lot of energy will be spent on yaw control—the job of which is to put the wind turbine in the direction of the wind. Having a variable direction may negate the advantage of higher wind speeds.

Obstacles: height and distance to turbine
The taller the obstacle, the larger the wind shades. If the turbine is closer to the obstacle than five times the obstacle height, or if the obstacle is taller than half the hub height, the results will be more uncertain because they will depend on the exact geometry of the obstacle. The obstacle, if it exists, should not be higher than the site.

The distance between the obstacle and the turbine is very important for the shelter effect. In general, the shelter effect will decrease as you move away from the obstacle, just like

(Continued)

SIZING OR MATCHING WINDMILLS TO SITES—cont'd

a smoke plume becomes diluted as you move away from a smokestack. If the turbine is closer to the obstacle than five times the obstacle height, the results will be more uncertain because they will depend on the exact geometry of the obstacle. Therefore, a site should not have many obstacles.

Surface roughness
The Earth's surface, with its vegetation and buildings, is the main factor in reducing wind speed. This is sometimes described as roughness of the terrain. As you move away from the Earth's surface, roughness decreases and the laminar flow of air increases. Expressed another way, increased height means greater wind speeds. High above ground level, at a height of about 1 kilometer, the wind is hardly influenced by the surface of the Earth.

In the lower layers of the atmosphere, however, wind speeds are affected by the friction against the surface of the Earth. For wind power utilization, this means that the higher the roughness of the Earth's surface, the more the wind will be slowed down. Wind speed is slowed down considerably by forests and large cities, while plains like water surfaces or airports will only slow the wind down a little. Buildings, forests, and other obstacles are not only reducing the wind speed but they also often create turbulence in their neighborhood. Water surfaces have the lowest influence on wind speed.

Access to existing utility transmission lines
The site should be close to the utility supply lines and the consumer. This will reduce the cost per kWh generated and will reduce losses if the distance to the customer is large. The site evaulation for locating a WECS should take this issue into consideration.

Potential legal and environmental obstacles

Look into the zoning issue. Are you allowed to put an energy conversion system at this site? The public's concern is rooted in the fact that the environmental advantages of wind power is on a global or national level, whereas the environmental disadvantages of wind power are on a local or neighborhood level and are specifically associated with the presence and operation of wind turbines.

Environmental Issues
Advantages on a global or national level. Some of the advantages are the following:

No direct atmospheric emissions: None are caused by the operation of wind turbines. The indirect emission from the energy used to produce, transport, and decommission a wind turbine depends on the type of primary energy used.

No liabilities after decommissioning: Electricity from wind turbines has no liabilities related to decommissioning of obsolete plants. Today, most wind turbines' metal parts can be recycled. In the very near future, other parts (e.g., electronics and blades) will able to be recycled almost 100%.

Limited use of land: Nearly 99% of the area occupied by a wind farm can be used for agriculture or remain as natural habitat. Furthermore, part of the installations can be made offshore. Consequently, limiting land area is not a physical constraint for wind power utilization.

Disadvantages on a local or neighborhood level. Some of the disadvantages are of very limited significance:

Noise emission: Acoustic emissions from wind turbines are composed of a mechanical and an aerodynamic component, both of which are a function of wind speed. Analysis shows that for most turbines with rotor diameters up to 20 m, the mechanical component dominates, whereas for larger rotors the aerodynamic component is decisive. The nuisance caused by turbine noise is one of the important limitations of sitting wind

SIZING OR MATCHING WINDMILLS TO SITES—cont'd

farms close to inhabited areas. The acceptable emission level strongly depends on local regulations

Visual impact on landscape: Depending on the characteristics of the landscape, modern wind turbines with a hub height of 40 to 60 m and a blade length of 20 to 30 m form a visual impact on the landscape. This visual impact, although very difficult to quantify, can be a planning restriction in most countries.

Moving shadows: A more objective case of visual impact is the effect of moving "shadows" from the rotor blades. This is only a problem in situations where turbines are sited very close to workplaces or dwellings. The effect can be easily predicted and avoided through proper planning. A house 300 m from a modern 600 kW machine with a rotor diameter of 40 m will be exposed to moving shadows ~17 to 18 hours out of 8760 hours annually.

Impact on birds. The impact of wind turbines on birds can be divided into:

- Direct impact, including risk of collision and effect on the breeding success.
- Indirect impact, including effects caused by disturbance from the wind turbine (noise and visual disturbance). The disturbance effects of wind turbines fall into three categories:
  - Disturbance to breeding birds
  - Disturbance to staging and foraging birds
  - Disturbing impact on migration/flying birds

Studies in Germany, the Netherlands, Denmark, and the United Kingdom have concluded that wind turbines do not pose any substantial threat to birds (or bats or insects). Bird mortality due to wind turbines is only a small fraction of the background mortality. One study estimated a maximum level of bird's collision with wind turbines of 6 to 7 birds/turbine/year. In Denmark, with approximately 4000 wind turbines, this means that 25,000 to 30,000 birds die annually from collisions with wind turbines. As a comparison, it can be mentioned that more than 1,000,000 birds are killed in traffic in Denmark, and that the total number of staging and migrating birds in Denmark is 400 to 500 million.

Isolated examples of significant damage have been reported to specific species—that is, in the vicinity of the Spanish wind farm of Tariff near the Strait of Gibraltar, which is a major bird migration route. If not properly dealt with, wind farms sited on coastal sites can disturb breeding and resting birds. Typically, an effect has been recorded within 250 to 800 m, with the highest sensitivity recorded for geese and waders. Including professional knowledge about birds and wind turbines in the planning process for wind farms can address this issue.

Interference with electromagnetic systems. Wind turbines in some areas can reflect electromagnetic waves, which will be scattered and diffracted. This means that turbines may interfere with telecommunication links. An investigation made by the British company BBC concluded that wind turbines' interference with electromagnetic communication systems is not a significant problem.

Personal safety

Accidents with wind turbines involving human beings are extremely rare, and there is no recorded case of persons hurt by parts of blades or ice loosened from turbines. US insurance companies, where most of the experience with large wind farms has been taken place, agree that the wind industry has a good safety profile compared to other energy-producing industries. The International Electrical Committee (IEC) has issued an international official standard for wind turbine safety.

3.8 WIND TURBINE SITE MATCHING

Given a chosen site, which wind turbine will be optimum for it? This is a job for someone with experience with all types of turbines. Not only must the wind turbine be well made, but it also must fit the wind conditions at the particular site and must produce the power that the system requires.

When choosing a wind turbine, the rated power for a it is not a good basis for comparing one product to the next. This is because manufacturers are free to pick the wind speed at which they rate their turbines. If the rated wind speeds are not the same, then comparing two products is very misleading. Usually, manufacturers give information on the annual energy output at various annual average wind speeds.

The selection of the optimum wind turbine for a specific site is done based on finding the capacity factor, CF, of the available wind turbines. The following are the procedures:

Generate a typical day with 24 hours of every month of the year: From the available long-term data, find the wind speeds at every hour of a typical day for every month. For a typical day in the month of June, for 8 o'clock in the morning, find all the speeds that ere recorded at that time of every day, for all the available June months from the long-term data.

Upgrade the data to the level of the hub: Upgrade the data to the level of the hub heights of every wind turbine that you may consider using for the site.

Generate mean wind speeds: Available long-term wind speed data for the site is recorded at different hours of the day for many years. Upgraded data is then used to generate mean wind speeds for every hour of a typical day for every month of the year.

Generate the probability density functions: The probability density function for the mean wind speeds for the different hours of the typical day are also generated.

Calculate the CF for every wind turbine being considered: The available wind turbines' manufacturing specifications, along with the mean wind speeds and the probability density functions, are used to calculate the capacity factors of all available turbines for every hour of the day. The CF is defined as the average power output of a wind turbine, $P_{w,\mathrm{avr}}$, divided by the rated output of a wind turbine, $P_{w,\mathrm{rated}}$,

$$\mathrm{CF} = P_{w,\mathrm{avr}} \div P_{w,\mathrm{rated}} \tag{3.30}$$

$$P_{w,\mathrm{avg}} = \int_{0}^{\infty} P_w \cdot f(v) \cdot \mathrm{d}v \tag{3.31}$$

Plus,

$$P_w = \left(\frac{1}{2}\right) \times \rho \times \left(V^3\right) \times A \times C_p \times \eta \qquad (3.32)$$

Find the average CF for every wind turbine considered: The wind turbine with the highest average capacity factor is the optimum type for the given site.

3.9 APPLICATIONS FOR WECS

Wind energy will be one of the most important, widely applied of the renewable energy forms during the next several decades. Successful research and development will potentially result in generation from wind energy of about 10% of the electricity used in the United States. The significant environmental and societal benefits of wind energy are limitless. Further advancements will find new applications for wind energy in the bright future.

Wind turbines are a crucial part of the movement toward using clean, renewable energy around the world. One of the great advantages of the wind turbine is that it can be used to provide energy to remote places that are far from the utility grid. Other advantages include reduction of utility bills, crop irrigation, pumping drinking water to nearby villages, and providing power to areas with harsh conditions.

WECSs could be used as the sole source of energy working in a *stand-alone mode* in the absence of grid transmission lines or a supplementary energy source. They also could be connected to a utility lines working in a *grid-connected mode*. They can be used on a small scale to power a house, an intermediate scale to power a village or island, or as a large-scale generating system for hundreds of megawatts for utility companies.

3.9.1 Stand-alone operation

In many years, wind power is the least costly for providing power to homes and businesses that are remote from an established grid. Researchers estimate that wind produces more power at less cost than diesel generators at any remote site, with an average wind speed greater than about 4 m/sec. Diesel generators, or photovoltaic systems, are used in addition to increase the reliability of stand-alone WECS operations.

3.9.2 Small-scale applications

The first application of stand-alone WECS is the small-scale application that includes supplying electricity to households, farms, and remote areas such as telecommunication relay stations, marine navigation aids, metrological stations, rail

FIGURE 3.45 A windmill system next to a remote home.

signaling systems, boats, and harsh places—for example, a monitoring post in the Arctic Circle. Plus, wind energy can be used to provide remote communities with clean drinking water. Figure 3.45 shows a remote house powered by a windmill.

Some villagers have to travel a few miles to get water and carry the heavy jugs back to their village. With the use of wind energy, it can be pumped from a water source to remote villages (Figure 3.46). Small wind power systems are ideal for

FIGURE 3.46 Community water supply in Niama District, Oujda, NE Morocco.

FIGURE 3.47 Remote weather system for the Finnis Coastguard.

applications where storing and shipping fuels is not economical or impossible. Wind turbines are used for powering remote weather stations; Figure 3.47 shows a wind turbine powering a Meteorological Station for Finnis. Coastguard.

Wind turbines can be used to irrigate farmland so that farmers can get more crops to grow on their land and feed livestock (Figure 3.48). Small wind systems have a higher initial cost but have a lower life-cycle cost than the kerosene pumps used for irrigation. In addition, using wind energy does not pollute the air and does not need the constant refueling and maintenance that the kerosene pump needs.

Wind turbines are used at telecommunication sites because they are usually remote—that is, far away from the utility grid. The high altitude of most communication sites provides a great resource of wind to be used by the turbines. Because of this there should be enough wind to power everything needed most of the time. A microwave repeater is located on Duncan Mountain in Idaho; a 7.5 kW wind turbine and a small solar array are all that is needed to power this site. After fifteen years of operation, the wind turbine is still providing almost all the energy required by the Idaho site.

Wind turbines can operate under extreme conditions, which allow them to be used when other energy sources would be impractical. In a place close to the Antarctica, temperatures range from –40°F in the winter to +40°F in the summer, so the energy generator must be quite robust to withstand these extreme temperature ranges. Wind turbines have been proven to withstand such conditions a lot better that other sources for generating electricity. Figure 3.49 shows an excellent use

FIGURE 3.48 A windmill for stocking water.

FIGURE 3.49 Assembly for a windmill in Antarctica.

FIGURE 3.50 A boat using wind power.

for tiny, lightweight turbines to keep emergency batteries charged or to provide electricity for boats. The electricity output ranges from about 150 to 500 watts (Figure 3.50).

Due to the terrain, there are many sites that can only be reached by air. Running utility lines to just a few places in remote locations over harsh terrain would be very impractical, so a remote source of energy is needed. Wind turbines are replacing the diesel generators that were used to provide energy to such sites. They need little maintenance, no refueling, and are a clean source of energy. Wind turbines will save the operators of the sites a lot of money because fuel will not have to be constantly flown in to power generators.

The size range of the small-scale WECS goes from a few watts to 20 kW. These stand-alone systems usually run with batteries to store the excess energy to be used at times when there is no wind. Some applications do not require batteries (e.g., water pumping for agriculture). The small-scale stand-alone systems usually have generators that could be DC ones that charge a battery directly and for a DC load or AC brushless synchronous generators.

3.9.3 Intermediate-scale application

The second application of a stand-alone WECS is the intermediate-scale of applications. These off-grid applications may include installing clusters of smaller wind turbines in the size range for 10 to 100 kW each in villages, on offshore islands, in isolated communities, and at enterprises that generate their own power.

FIGURE 3.51 Ten wind turbines powering a remote village in Kotzebue, Alaska.

Many turbines are installed for rural electrification of villages and to reduce the use of expensive and heavily polluting diesel-driven generators in areas that are beyond the national electricity grid system. North of the Artic Circle in Kotzebue, Alaska, wind turbines generate lower-cost electricity for the community of 3500 residents, while reducing environmental risk from transport, storage, and burning of diesel fuel. The system was installed in April of 1999. During the February reporting period, the project generated 56 MWh, which equates to a 38% CF assuming, a sustained peak power rating of 66 kWh per turbine (Figure 3.51).

It is very costly and dangerous to run power cables across or under the water to a location that needs electricity. It would be much less expensive and safer to install a system onsite that would generate power. Wind energy is an excellent choice because the turbines would be located close to the coastline. Wind is created on the shore due to the way the Sun heats the land and the ocean differently. This difference in temperature moves the air, creating wind. This makes the coastline an area with an abundance of wind resources that can be harnessed by turbines.

Islands have difficulty getting electricity because using underground cables is quite expensive and impractical if it is far away from the mainland. Areas with many little islands would be best off using wind energy or a hybrid of wind and solar energy. So, instead of trying to tie all islands together with an underground cable that carries electricity, each island would have one or more wind turbines to provide energy for it.

FIGURE 3.52 Wind/PV hybrid system for powering the Caribbean Islands.

An island in the Caribbean is powered by a hybrid wind/PV system is shown in Figure 3.52. The wind turbine serves as a primary energy generator and solar panels serve as a secondary energy source. A diesel generator acts as a backup during days when there is no wind or sun. The villager is able to enjoy grid-quality, 24-hour AC power. Xiao Qing Dao Island Village in China is powered by a wind/diesel hybrid system installed in February 2001. The wind turbine type is the Bergey (Figure 3.53).

Krasnoe Island in Russia, at the mouth of the Devina River, is one of 21 pilot village electrification projects throughout Northern Russia. There are approximately 20 million people living without access to the electricity supply gird. The system was installed in September 1997. It consists of two 7.5 kW turbines (Figure 3.54) on 24-m tilt up towers, a 48 VDC battery bank, three 4.5 kW inverters, and associated switching gear. The system works with the existing village diesel generator to provide 24-hour power with minimum diesel fuel consumption. The site conditions in the winter have temperatures reaching –40°F.

3.9.4 Utility grid-connected WECS

In this mode of operation the WECS is connected in parallel with an existing grid. Again it could be a small-scale application, where a windmill is powering a house or a business; or an intermediate-scale application, where a cluster of wind turbines are used to power a village or an industrial area; or a large-scale application, where multi-megawatt power is generated from a wind farm formed of multiple turbines.

Homes and businesses that are located in windy areas and have fairly high utility bills should consider purchasing a wind turbine to provide them with clean power

FIGURE 3.53 Bergey Excel 10 kW wind turbine in a Xiao Qing Dao Island village.

FIGURE 3.54 Village electrification system for Krasnoe Island, Russia.

without relying solely on the utility grid. The initial cost is high, but the savings will pay it off after about 7 to 10 years. Extra electricity that the turbine produces can be sold to the utility company.

The amount the utility pays for that electricity varies by the area. Some places will spin the meter backward when extra electricity is being produced. This is the best situation because you are selling the electricity for the same amount you pay for it. If this is the case, then it is quite possible for the owner to not pay anything for electricity because the amount of electricity taken from the utility grid equals the extra amount produced from the wind turbine over the entire year.

3.9.5 Wind farms

Wind farms are clusters (arrays) of multiple wind turbines. The number of turbines in the array varies from two to three in a small cluster to the thousands of machines in California's windy passes. Concerns about the impacts that traditional power-generating methods (i.e., fossil fuels) have on the environment, high gas prices, and the fact that many of our nonrenewable resources are quickly depleting have led to the belief that radical rethinking is now needed as to how energy of the future gets generated. These concerns have caused an increased interest in so-called "environmentally friendly" renewable energy sources. Sources include those that occur naturally and repeatedly in the environment such as energy from the Sun, the wind, the oceans, the Earth, waste (biomass), and the fall of water.

Wind power is an energy source that has been used in some capacity for hundreds of years. But not until the last few decades was it that the largely untapped treasure of wind was considered as an alternate source of US energy that was then brought to life in the form of turbines in wind farms across the nation. Wind power is one form of renewable energy that is on the verge of widespread commercial exploitation. Under the right conditions, wind energy can now produce less expensive electricity than both coal and nuclear power stations and is equivalent to gas-fired power stations.

In June 1999, the US government announced the Wind Powering America Initiative to build on current and future public and private-sector efforts to support the development of wind power. The following are the three main goals of this initiative:

- Supply at least 5% of the nation's electricity from wind energy by the year 2020, creating 80,000 jobs and displacing 35 million tons of atmospheric carbon, with more than 5000 MW of wind energy installed by 2005 and 10,000 MW online by 2010
- Double the number of states that have more than 20 MW of wind capacity to 16 by 2005 and triple the number to 24 by 2010.
- Increase the federal government's use of wind-generated electricity to 5% of electricity consumption by 2010. Utilities or electricity suppliers in every state may now offer or plan to offer green power produced from wind and other renewable energy resources (Wind Energy, 1999).

The central region of the United States has the potential to be the primary producer of wind power. Every state in the nation has the resources to harness some wind energy, but states such as North Dakota, which has an abundance of resources that could supply 36% of the electricity consumed in the 48 contiguous states, are the "oil wells" of wind power. The amount of energy is measured in kilowatt-hours—the term describes the wind power density in watts per square meter, which accounts for both wind speed and the distribution of wind over time.

There are many criteria that can be used as determinants for selecting sites that may be suitable for a new wind farm development. The relative importance of each is very dependent on the viewpoint of the user (i.e. environmentalist, conservationist, or developer). However, if the criteria themselves are subjective, then the limits associated with each are even more so. A major economic concern for utilities is the distance from the proposed wind farm to existing high-voltage power lines.

Construction of high-voltage lines is very expensive. As a general rule, wind farm sites cannot be more than 10 miles from an existing high-voltage power line. It may also be obvious that a wind farm should not be too close to an airport. It is imperative to know what the actual wind speed is at a site before developing a wind farm. Three years of wind data are usually considered the minimum requirement. In addition to wind speed, there are economic and environmental considerations when choosing a wind farm site.

The large wind farms have been installed in California and there are many reasons for this. Wind farms there are in places where very favorable winds occur. They are also near electric power transmission lines and large cities. Peak winds in these areas occur approximately at the same time as the peak electricity demand. This increases the value of the electricity to consumers. Another important factor is that the state of California requires electric utilities to purchase the electricity from the wind farms at favorable prices to the turbine owners. There is a large wind farm at Altamont Pass, California, just east of San Francisco (Figure 3.55).

FIGURE 3.55 The Altamont Pass wind farm.

FIGURE 3.56 The Texas Horse Hollow Wind farm produces 735 MW of power.

Another economic concern is the cost of constructing a wind farm. Is the site already accessible by road or will a road need to be built? What is the slope of the site? Steep sites require extra construction expense. Other economic concerns include: What are the property taxes? How long will it take for the wind farm to make a profit? The environmental impact of a wind farm is also a concern. Some people are concerned because of the large land area required for the turbines; large installations may require hundreds of acres. However, only 5% of the land of a wind farm is needed for towers, roads, and support structures. The remaining 95% of the land may be used for other purposes such as row crops or ranching.

The largest wind farm was constructed in Texas (Figure 3.56); it is called the "Texas Horse Hollow Wind Farm" with a capacity of 735 MW. The windmills are erected in rows that need to be located at certain distance from each other so that they do not fall in each others shadow. Increasing the distance between the rows decreases the number that can be installed in a certain area; on the other hand, decreasing the distance will cause the wind turbines to operate in the wake of those in the front, leading to disturbance and weakening of the wind speed. Wake, by definition, is the region around the windmill where the pattern of the wind is disturbed. This eventually will lead to reducing the capacity factor and therefore the amount of energy generated. A rule of thumb is that a wind turbine should be located at a distance 10 to 15 times its diameter downwind and cross wind. Denmark is the country that uses the most WECSs; it has more than 4000 wind turbines. Figure 3.57 shows an offshore wind farm.

As cost effective and environmentally safe as it is, wind power is not without its complications. The most common complaints, according to the US Department of Energy, are the noise pollution created by the rotors, the lack of aesthetic appeal,

FIGURE 3.57 Offshore wind farm in Denmark.

and the risk of bird mortality because they fly into the rotors. In California, birds of prey (raptors) have been killed by windmills; current research indicates that this may be a local problem and seems to be associated with fog. The National Renewable Energy Laboratory is conducting studies to address the issue of avian mortality in order to prevent both a decline in bird populations and violations of the Migratory Bird Treaty Act or the Endangered Species Act.

As for the noise generated by the wind turbines, early generation ones were very noisy. The newer turbines are much quieter, barely audible at a distance of 100 yards. A final environmental concern is aesthetics—the look of wind farms. On most occasions, the turbines are sited in remote locations, so the noise and aesthetics do not affect large numbers of people; regardless, some people still consider them a blight on the landscape. Others find their turning in the wind graceful.

3.10 LEVELS OF WIND POWER GENERATION IN UTILITY GRID SYSTEMS

Grid integration concerns have come to the forefront in recent years, as wind power penetration levels have increased in a number of countries, as an issue that may impede the widespread deployment of wind power systems. Two of the strongest challenges to wind power's future prospects are the problems of intermittency and grid reliability.

Electricity systems must supply power in close balance to demand. The average load varies in predictable daily and seasonal patterns, but there is an unpredictable component due to random load variations and unforeseen events. To compensate for these variations, additional generation capacity is needed to provide regulation or set aside as reserves. Generators within an electrical system have varying operating characteristics: some are base-load plants; others (e.g., hydro or combustion turbines) are more agile in terms of response to fluctuations and startup times. There is an economic value above the energy produced to a generator that can provide these ancillary services. Introducing wind generation can increase the regulation burden and need for reserves due to its natural intermittency.

The impact of the wind farm's variability may range from negligible to significant depending on the level of penetration and intermittency of the wind resource. The intermittent profile can affect grid reliability and is a new dimension that transmission and distribution network operators have not traditionally had to manage on any significant scale. Consequently, as wind power has become more than a novelty in electricity supply, system planners and operators have been increasingly concerned that variations in wind turbines' output may adversely affect grid reliability and increase the operating costs of the system as a whole.

The conventional management of transmission and distribution operation is challenged by electricity market restructuring, security of supply concerns, and the integration of newer generation technologies such as wind power. Yet some developments of market restructuring, such as increasing network interconnections, can improve grid integration and management of variable generation sources. It can provide opportunities for market designs that reduce costs for balancing power.

In some countries, the transmission and distribution networks are aging or are at capacity. Infrastructure replacement and upgrading is an opportunity to reexamine design and operation parameters including integrating wind power generation as a key part of the overall network strategy. This will require system operators to develop active network management—a radical shift from the traditional central control approach.

Offshore wind development presents a lower level of technological maturity and higher risks than onshore wind power developments. While R&D advances and market learning have been observed in the areas of turbines, base structures, operation and maintenance, and communication relays for offshore development, all need further attention. Then too, developments offshore will entail transmission line construction.

In many cases, it is expected to be high-voltage direct current (HVDC) technology, which comes at increased expense but offers additional functionality for grid integration and low line losses. Offshore turbines are expected to be in the multi-megawatt range for development in the near term, thereby bringing larger amounts of intermittent resources into the generation mix and increasing the need to find solutions to the integration challenges.

Transmission availability can be a barrier to wind power development. Favorable wind locations are often in areas distant from existing transmission facilities—for example, offshore areas in Europe and rural areas in North America. Building new transmission lines can be difficult due to planning barriers, land use rights, and costs.

Due to the precariousness associated with wind power integration into the utility system, there are many theories about what credit to give the wind power, or what percentage of the total utility system capacity should be provided by wind. The following four theories need to be considered:

Fuel saver, 0%: In some cases, it is assumed that the credited capacity of wind power is zero; that is, wind power is so unreliable that, for all practical purposes, no decrease in the conventional capacity otherwise needed is possible. The wind-generated energy is then said to contribute as a "fuel saver" such that its contribution to the utility is measured only by the fuel not consumed when wind power is available. The economic value of wind turbines is then computed as the cost of this "saved fuel." The advantage of this approach is that it is simple to implement, but it can result in an overly conservative (and thus expensive) commitment of part-loaded generating units. In the case of combined cycle gas turbines (CCGTs), which are supplying an ever-increasing fraction of energy demands, this can be particularly significant as oxides of nitrogen (NOx) emissions rise dramatically when such a plant is part-loaded.

Hot spinning reserve: This theory states that wind power should be given a percentage that is equal to the utility spinning reserve. These generators will pick up the load when the wind dies out, thus the utility will not suffer. Utilities carry operating reserves to ensure adequate system performance and to guard against sudden loss of generation, off-system purchases, unexpected load fluctuations, and/or unexpected transmission line outages. Operating reserve is further defined to be spinning or nonspinning reserve. Typically, one-half of system operating reserves is spinning so that a sudden loss of generation will not result in a loss of load, with the balance available to serve load within 10 minutes. Any probable load or generation variations that cannot be forecast have to be considered when determining the amount of operating reserve to carry.

At current wind plant penetration levels in California, the variability of wind output has not required any change in operating reserve requirements. The exact point at which the integration of intermittent generation, such as wind, begins to degrade system economics is unclear, but the technical literature suggests that it is at penetration levels in excess of 5%. Intermittency is becoming an increasing concern to utility operators in California, particularly during low-demand periods since wind farm penetration is beginning to reach this level. As markets for electricity become more competitive, the ability to forecast and control the wind resource will increase the value of wind energy to utilities.

Peaking units capacity: This theory stipulates that wind power should be given a percentage that is equal to the capacity of the peaking units. If the wind dies out, the peaking units will pick up the load. Electrical reliability is a function of customer demand and the various generator characteristics. Utilities experience a pronounced peak period, often several hours of the day during a particular season. A utility can sometimes dramatically increase its generator reliability by installing peaking units to generate power when it is most needed, during peak hours. Even though these peaking units might be available at night, their availability then would likely have a negligible affect on system reliability. Likewise, as we have already seen, a wind plant that delivers a significantly higher annual energy output does not necessarily contribute significantly to system reliability. What is needed is for wind output to occur at times of otherwise high-risk periods during system peak.

Capacity credit (CC): This is defined as the amount of conventional capacity to be installed that can be avoided because of the presence of a WECS. Wind energy is often criticized for being unreliable. Critics claim that wind energy can never replace existing power stations, or remove the need for new power stations to be built, because the wind cannot be relied on. In technical terms, this argument boils down to the question: Can wind energy be regarded as having a capacity credit? Wind energy can be relied on even though the wind is not available 100% of the time. In fact, no energy technology can be reliable 100% of the time. The concept of load factor deals with the day-to-day productivity of an electricity-generating plant.

No individual power plant is always available to supply electricity. All plants are unavailable at certain times, whether for routine maintenance or for unexpected reasons. The load factor of an energy technology is the ratio (expressed as a percentage) of the net amount of electricity generated by a power plant to the net amount that it could have generated if it were operating at its net output capacity. Wind farms can be treated statistically in exactly the same way as conventional power plants. For any type of them, it is possible to calculate the probability of it not being able to supply the expected load. Because the wind is variable, the probability that it will not be available at any particular time is higher. Wind energy has a lower load factor than many other technologies, as shown in the Table 3.1.

Table 3.1 Other Technologies and Load Factors	
Energy technology	**Load factor**
Sewage gas	90%
Farmyard waste	90%
Energy crops	85%
Landfill gas	70-90%
Combined cycle gas turbine	70-85%
Waste combustion	60-90%
Coal	65-85%
Nuclear power	65-85%
Hydro	30-50%
Wind enegy	25-40%
Wave power	25%

The load factor of wind varies according to the site and the type of turbine, but it is generally around 30% and is higher during the winter than the summer. An average wind farm with an installed capacity of say 5 MW will produce an output of 13,140 MW hours/year—that is, 30% of what it would produce if it were operating continually at maximum output.

Capacity credit is calculated by determining the reductions of installed power capacity at thermal power stations so that the probability of loss of load at winter peaks is not increased. Or, in other words, how much of the thermal power plant could be "replaced" by wind power without making the system less reliable. For low levels of wind power penetration into the grid, the capacity credit of wind energy is about the same as the installed capacity multiplied by the load factor. In other words, if there is 100 MW of wind energy installed in the country, then this can be relied on to replace (or avoid the need to build) 30 MW of thermal- or nuclear-generation capacity. However, as the level of wind penetration rises, the capacity credit begins to tail off.

To find the capacity credit for a region with a given system installed generating capacity (SIGC) and wind speed data, the following steps should be taken:

1. Calculate the loss of load probability for different scenarios of loads without the WECS. Each generator has a forced output rate of 5% for coal-powered generators, 1.5% for hydro generators, and so on. The energy generated by all generators is calculated considering the outages of all of them. The loss of load probability (LOLP) is defined as the probability that the load exceeds the SIGC. The equivalent load duration curve (ELDC) is then constructed. The reliability is defined as the probability that the SIGC is more than the load. Of course if the load is zero, then the probability that the SIGC is more than the load is 100% or one, no risk.
2. Repeat Step 1 considering the group of wind turbine generators' capacity as negative loads.
3. To find the capacity credit of a group of generators for this region, you must decide the risk you are going to take and draw a horizontal line. The line will intersect the two curves. The difference between the two corresponding capacities is the capacity credit.

$$CC = SIGC - 1 - SIGC - 2$$

The more risk you take the higher the capacity credit WECS will assume (Figure 3.58). Wind energy does have a capacity credit and can therefore be relied upon, although the wind isn't always available. Capacity credit falls as penetration of wind and other nonfirm technologies that go into the system increases; this will not be an issue until levels reach approximately 20% penetration. The load factor is sometimes called the capacity factor. Moreover, in a region where wind does not die out, sometimes with a small WECS, the capacity credit is taken equal to the capacity factor.

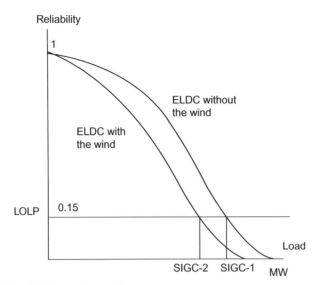

FIGURE 3.58 The WECS capacity credit.

3.11 FUTURE EXPECTATION OF WECS

The future for WECS is very bright, and the cost is constantly going down due to the mass production, innovation, and the invention of new products. The world's largest wind turbine is produced by Germany (Enercon E-126); the capacity is 6 MW (Figure 3.59). The worldwide production of wind energy is increasing very rapidly as can be seen from Figure 3.60.

FIGURE 3.59 The world's largest wind turbine is built in Germany.

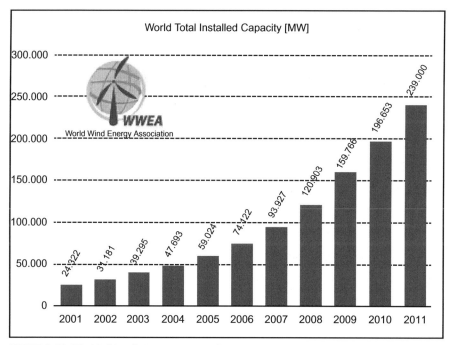

FIGURE 3.60 Worldwide wind power production.

FEDERAL WIND ENERGY PROGRAM (1972–1984)
Government-sponsored R&D
The federal wind energy program was created to commercialize wind power by:

- Supporting research and technology development for wind turbines.
- Design, build, and operate a WECS for research and development. This was carried out by the Department of Energy.
- Initiate studies to determine WECSs' viability for utility applications.

The DOE developed and built a series of WECS over the life of the program.

MOD-0
This was the first WECS in the program (Figure 3.61); it was built by Westinghouse in 1975. Here are its specifications:

- Variable pitch–angle control
- Horizontal axis machine
- Built in Ohio in Plumbrook.
- Wind turbine diameter: $D = 38$ m
- Rated power output: $P = 100$ kW
- Number of Blades: 2
- Cut-off wind speed: $V_{wo} = 17.9$ m/s
- Cut-in wind speed: $V_{wi} = 4.3$ m/s
- Rated wind speed: $V_{wr} = 10.0$ m/s
- Swept area by the wind turbine: $A = 1072$ m^2

FEDERAL WIND ENERGY PROGRAM (1972–1984)—cont'd

FIGURE 3.61 The MOD-0 wind turbine.

- Wind turbine power coefficient: C_p max $= 0.375$
- Hub height: 30 m
- Hybrid type: rigid
- Weigh of the two blades: 2090 kg
- Rotor orientation: Downwind
- Control type: Full span
- Generator type: Regular synchronous
- Generator voltage: 480 V

MOD-OA
This one was also built by Westinghouse in Clanton, New Mexico, in 1979. Four machines were built (Figure 3.62), as follows:

- Turbine diameter: $D = 37.5$ m
- Rated output power: $P = 200$ kW
- Cut-off wind speed: $V_{wo} = 17.9$ m/s
- Cut-in wind speed: $V_{wi} = 5.4$ m/s
- Rated wind speed: $V_{wr} = 9.7$ m/s
- Sweep area: $A = 1140$ m^2

(Continued)

FEDERAL WIND ENERGY PROGRAM (1972–1984)—cont'd

FIGURE 3.62 The MOD-OA wind turbine.

- Power coefficient: C_p max $= 0.375$
- Maximum wind speed: 67 m/s
- Hub height: 30 m
- Generator type: Regular synchronous generator
- Generator voltage: 480 V
- Tower construction: Tubular steel

MOD-1
Built by General Electric in Boone, North Carolina, with variable pitch–angle control (Figure 3.63); it is a horizontal axis machine with these specifications.

- Turbine diameter: $D = 60$ m
- Rated output power: $P = 2$ MW
- Cut-off wind speed: $V_{wo} = 19$ m/s
- Cut-in wind speed: $V_{wi} = 7$ m/s
- Rated wind speed: $V_{wr} = 10.0$ m/s
- Sweep area $= 2920$ m^2
- Power coefficient: C_p max $= 0.375$
- Weight of the two blades: 16,400 kg
- Downwind
- Full spam

FEDERAL WIND ENERGY PROGRAM (1972–1984)—cont'd

FIGURE 3.63 NASA MOD-1 wind turbine.

- Generator type: Regular synchronous
- Generator voltage: 4160 V
- Hub: rigid
- Generated TV interference, noise

MOD-2
This one was built by Boeing and installed in Goldengale, Washington State, in 1981 (Figures 3.64 and 3.65). The following are its specifications:

- Turbine diameter: $D = 91$ m
- Rated output power: $P = 2.5$ MW
- Cut-off wind speed: $V_{wo} = 20.1$ m/s
- Cut-in wind speed: $V_{wi} = 6.3$ m/s
- Rated wind speed: $V_{wr} = 10.0$ m/s
- Sweep area $= 6560$ m^2
- Power coefficient: C_p max $= 0.382$
- Weight of the two blades: 33200 kg
- Steel blades
- Upwind
- Generator voltage: 4160 V
- Hub height: 58.8 m
- Partial pitching
- Generator type: Regular synchronous

(Continued)

FEDERAL WIND ENERGY PROGRAM (1972–1984)—cont'd

FIGURE 3.64 A MOD-2 wind turbine.

FIGURE 3.65 A farm of three MOD-2 wind turbines.

FEDERAL WIND ENERGY PROGRAM (1972–1984)—cont'd

FIGURE 3.66 A MOD-5B wind turbine.

MOD-5A

It was designed by General Electric in 1984 in Hawaii; however, it was never built.

- Turbine diameter: $D = 122$ m
- Rated output power: $P = 7.3$ MW
- Rated wind speed: $V_{wr} = 13$ m/s
- Wood rotor
- Three-bladed turbine
- Two rotational speeds: 13 and 17 rpm
- Projected cost: 3.75 c/kWh (currently 16 c/kWh)

MOD-5B

This unit was designed by Boeing (Figure 3.66) with these specifications.

- Turbine diameter: $D = 128$ m
- Rated output power: $P = 7.2$ MW
- Steel blades with wooden tips
- Generator: Variable speed

3.12 A HYBRID WIND AND PV ENERGY CONVERSION SYSTEM

A growing interest in renewable energy resources has been observed for several years. The alternative energy sources are nonpolluting, free in their availability, and continuous. These facts make them attractive for many applications.

3.12.1 Description

Today, hybrid wind and PV systems (HWPS) are used in various applications such as water pumping, lighting, electrification of remote areas, and telecommunications. For remote systems like radio telecommunication, satellite earth stations, or at sites that are far away from a conventional power system, the HWPS is considered preferrable. As the demand for viable, renewable energy-generating systems has increased, the two most widely used of these sources have been solar energy and wind energy.

When used in isolation as sources for distributed energy generation individually, each of the resources exhibits some significant drawbacks. One popular solution has been the creation of systems that utilize both solar energy and wind energy in tandem, often described as hybrid systems. This chapter discusses the difficulties inherent in isolated wind and photovoltaic systems and examines how hybrid wind and PV systems can combat these flaws. Drawing on recent research, it also sites examples of HWPS used in a wide variety of settings (Figure 3.67).

Photovoltaic systems have been a mainstay of the movement toward renewable energy since the earliest days. They are modular, allowing the owner to add onto the system incrementally as need arises or funds become available. While the level of irradiance and insolation does vary with location, season, and time of day, sunlight is nonetheless one of the few sources of energy that could conceivably be tapped virtually anywhere on the planet.

The obvious challenge faced by designers of stand-alone PV systems, however, is the fact that a given location on Earth receives sunlight for only a limited time each day before night intervenes. In many locations, the onset of winter or simply periods of cloudiness also limit the availability of sunlight. Since the load demand does not drop off at coinciding intervals, an energy-storage system (usually batteries) must be incorporated into the design to capture the excess energy generated during peak times for use during lulls. The need for a large storage capacity increases the cost of the system considerably. Often, reliance on a backup power generator or external grid is also necessary to ensure constant availability, which means higher costs as well.

Like sunlight, wind is a free and nonpolluting energy resource, although for some of the larger turbine models noise pollution must be taken into account; that too, can be modular. Wind can be found in some capacity everywhere, even though its availability varies dramatically based on terrain and other environmental characteristics.

Stand-alone wind systems, however, have challenges similar to those of stand-alone photovoltaic systems. Wind speeds tend to fluctuate significantly from one

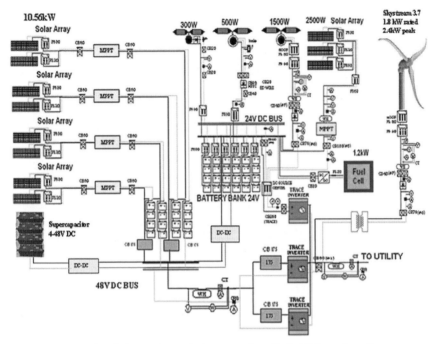

FIGURE 3.67 A hybrid wind and photovoltaic system installed at UMass–Lowell.

hour to another and from one season to another. Because of these frequent (and sometimes unpredictable) lapses in energy collection, a stand-alone WECS of a size appropriate for distributed generation does not produce usable energy for a considerable portion of time throughout the year and cannot satisfy constant load demands. As with PV power, considerable battery storage and reliance on an additional backup power source are necessary in order to ensure uninterrupted power, adding once again to the cost and complexity of the system.

A network that integrates both solar and wind power into one hybrid generation system has considerable advantages over its stand-alone counterparts. Most significantly, a mixture of wind and solar power can attenuate the individual variations in each source, both in terms of predictable weather patterns and less predictable fluctuations. This increases overall energy output and reliability and reduces energy-storage requirements and their associated costs.

3.12.2 Advantages of the HWPS

Irradiance and wind speed have different, predictable average peak and ebb points, both on a daily basis and on an annual basis. Figure 3.68 illustrates wind and solar energy generation for a typical day in the Merrimack River Valley of Massachusetts. In a daily cycle, solar energy generation tends to peak between noon and 3:00 pm,

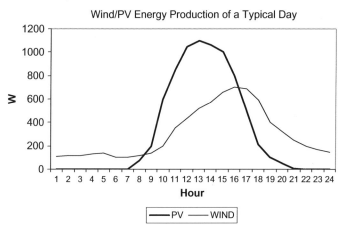

FIGURE 3.68 Wind and PV power production on a typical day in the Merrimack River Valley, MA.

while wind energy generation tends to peak between 3:00 pm and 6:00 pm. Solar energy peaks can accommodate midday loads such as air conditioning and commercial use, while wind peaks accommodate late afternoon commercial use and early evening residential demands. Only 10% of wind energy and little to no solar energy are generated between midnight and 6:00 am, which coincides nicely with the lull in load demand.

As can be seen from the curve in the figure, area 1 indicates low power production of both wind and PV systems. In this area, the need for battery power is clear. In area 2, a lot of energy is generated; the excess energy should be stored in a battery bank. Area 3 shows a close balance between the energy production and the load energy demand.

On an annual basis, the collection of wind energy peaks in the fall and winter months, coinciding with the winter peak load, while solar energy peaks in the spring and summer months, coinciding with the summer peak load. This seasonal variation suggests that wind power is most appropriate for heating and lighting applications that peak in the winter, while summer air conditioning and irrigation needs are best met by PV power. In this way, wind and photovoltaic together create a complementary system, extending the overall peak generation periods, both daily and annually.

The fact that these two forms of energy generation rely on two different natural phenomena also substantially increases the overall *reliability* of a system by reducing the negative impact of random weather incidences on it. While this is a beneficial feature of HWPS regardless of location, it contributes quite substantially to the advantage of such a system in an area where wind and/or sunshine are significantly unpredictable (e.g., Massachusetts's Merrimack River Valley).

Wind is more dynamic than the Sun and can produce energy when the sky is dark or overcast, while the Sun may shine brightly during times of low wind activity. This complementary combination of sources shrinks battery bank requirements because it means there will be fewer occurrences of emergency "downtime" when no power is being generated and the user has to draw on reserves. It also extends the life of the batteries by protecting them against deep discharges.

Although the overall amount of energy generation from a hybrid system may be slightly less than the hybrid components' individual capabilities if operated as larger, single-type systems, in most cases this is more than compensated for by the drastic reduction in expense. The cost of energy ($/kWh) will vary from site to site, since it is largely dependent on a number of site-related factors, including average wind speed, irradiance, and availability of assistance programs, governmental or otherwise.

Individually, the average costs of solar ($\sim$$0.065/kWh) and wind ($\sim$ $0.05/kWh) power generation, while still high compared to conventional, nonrenewable energy generation, are currently dramatically lower than in previous years. These prices are constantly decreasing because of the massive production of PV modules and small-scale wind turbines, progress in research and development, and continuous government financial support. This trend should continue to encourage the widespread application of interactive small-scale residential HWPS.

As with any distributed-generation system, the role of the battery in a HWPS is primarily to allow for the storing of excess energy produced at peak times for use during a time of little or no production. If the system is grid-connected, however, the excess energy may be sold back to the utility, providing additional economic incentive. In addition, storage can also help reduce the utility's peak-hour requirements by storing energy from it in time of low demand for use when the residential needs are at peak. Therefore, this reduces the scheduling of expensive generation units at the time of peak demand.

A HWPS is adaptable to a wide variety of settings. The most apparent of potential environments is one that boasts both high wind speeds, such as an island or coastal area, and high irradiance levels (e.g., an open desert or tropical region). The hybrid system is ideally suited for villages on small isolated islands because it will provide much-needed services without the exorbitant expense of building a connection to a far-away power grid, while making use of the village's abundant renewable, natural resources.

Remote, rural areas are, however, hardly the only place where a HWPS can be effectively used. An increasing number of hybrid systems are being developed as supplementary power sources for urban settings as well. In urban areas, where land is costly and people often oppose the construction of new power plants, a distributed generation system consisting of small-scale residential renewable energy hybrid systems can be integrated into the existing infrastructure to meet an increasing load demand with minimal ecological and societal impact. Such a distributed generation system can reduce system losses and be used as demand-side management to more closely match load with supply.

A good example of such a system is the HWPS currently in use on UMass–Lowell's campus (see Figure 3.67). This network consists of three wind turbines rated at 1.5 kW, 500 W, and 300 W and a 2500 W array of solar panels connected to a 24 V battery bank, as well as an inverter and a breaker. In general, smaller wind turbines fit well into an urban environment because they have a low cut-in speed, which in fact makes them better able to generate power in areas of overall low wind speed than their larger counterparts.

Their size also means a lower level of noise pollution, which will generally be drown out by traffic and other typical urban sounds; the environmental impact on the urban setting is negligible. Installation costs may be higher than in more rural settings because of the structures involved, and neighboring structures may hinder sun and wind resources somewhat, although wind speeds will be higher than they would be on the ground due to the elevation of the buildings.

Finally, HWPS technology is increasingly moving into the suburban landscape via private households and business. Many of these distributed-generation systems are used to reduce electric bills; the average US household spends close to $1400 per year for electricity.

Small-scale HWPS have been categorized as "the trend of the future" in a deregulated environment. They have been growing steadily in popularity and in refinement and have come to be seen as the most pragmatic approach to distributed energy generation in a diverse variety of settings. This consensus can only result in a positive trend toward distributed energy generation, a movement that helps to minimize environmental pollution, reduce losses in power systems' transmission and distribution equipment, and support the utility in demand side management (DSM). This results in increased reliability because a large geographical area becomes less dependent on a few distribution lines.

Other, less tangible incentives for investing in hybrid wind and photovoltaic distributed-generation systems are the ability of the customer to independently satisfy his or her energy needs in case of a blackout or other emergency, the long-term health benefits associated with pollution reduction, and a decrease of dependence on dwindling, nonrenewable resources, both foreign and domestic. According to the Synergy Power Corporation, a decreased dependence on foreign energy may also mean a decline in military expenditures to protect those sources and a less volatile energy market.

In conclusion, utilization of HWPSs is an excellent way to advance distributed energy generation. Solar and wind power sources complement one another in such a way as to extend the overall peak power-generation periods, conserving battery life and reducing the overall cost of the system. hybrid wind and PV systems are adaptable to a wide variety of environments, from rural villages to urban jungles, and are fast becoming the system of choice for distributed generation of renewable energy.

PROBLEMS

3.1 The wind speed at the height of 10 m is 8 m/sec. Calculate the wind speed at a height of 40 m for the following.
  a) The location is on land, $\alpha = 0.14$
  b) The location is on sea, $\alpha = 0.1$

3.2 In a location, 7 wind speeds were observed: 2, 3, 5, 7, 9, 11, and 12 m/sec. Calculate the V_{wm}, δ^2 and δ.

3.3 On a site, the number of wind speed observations is $n = 250$; each wind speed, V_{wi}, is observed m_i times. Calculate $P(V_{wi})$ and $F(V_{wi})$. Show the results in the table.

Item	Wind speeds	V_{wi} m/sec	Number of observations m_i	$P(V_{wi})$	$F(V_{wi})$
1	2	10			
2	4	20			
3	5	35			
4	7	60			
5	9	50			
6	10	40			
7	12	35			

3.4 A wind energy system has the following parameters:
Rotor turbine diameter: $D = 60$ m
Wind turbine maximum power coefficient: $C_{po} = 0.375$
Rated wind speed: $V_{wr} = 10$ m/sec
Cut-in wind speed: $V_{wi} = 7$ m/sec
Cut-off wind speed: $V_{wo} = 19$ m/sec
Air density: $\rho = 1.225$ kg/m^3
Synchronous generator rated power output: $P_o = 200$ kW
a) Calculate the available power in the wind at rated wind speed.
b) If the wind turbine operates at maximum power coefficient at rated wind speed, then calculate the mechanical power extracted from the wind at rated wind speed, P_m.
c) Calculate the combined gear generator efficiency, η.
d) Calculate the generator capacity factor, CF, if the average power output of the generator is $P_{av} = 400$ kW.

3.5 A wind energy conversion system has the following parameters:
Swept area by the blades of the turbine: $A = 6560$ m^2
Wind turbine maximum power coefficient: $C_{po} = 0.382$
Rated wind speed: $V_{wr} = 10$ m/sec
Cut-off wind speed: $V_{wo} = 20.1$ m/sec
Cut-in wind speed: $V_{wi} = 6.3$ m/sec
Air density: $\rho = 1.225$ kg/m^3
Synchronous generator rated power output: $P_o = 1500$ kW
a) Calculate the available power in the wind at rated wind speed.
b) If the wind turbine operates at a maximum power coefficient at rated wind speed, then calculate the mechanical power extracted from the wind at rated wind speed, P_m.
c) Calculate the combined gear generator efficiency, η.
d) Calculate the generator CF if the average power output of the generator is $P_{av} = 500$ kW.

3.6 A 5-kW DC generator is driven by a fixed pitch–angle wind turbine.
The generator delivers its rated output at a rated wind speed of 12 m/sec.

The generator is connected to a large 120 V battery bank. The generator is driven at rated wind speed, delivering its rated output.

a) Calculate the generator output current, I.

b) The wind speed was reduced to 8 m/sec; the generator efficiency stayed the same but the aerodynamic coefficient became 80% of what it was. Calculate the power output of the generator. The battery voltage stays constant at 120 V.

3.7 The equivalent load duration curve (ELDC) of a power system is given without wind and with wind energy as shown in the Figure 3.69, calculate the capacity credit of the wind energy conversion system for a loss of load probability, LOLP = 0.2.

3.8 For a certain wind site, the average wind speed is 6 m/sec and the variance is $\delta^2 = 8.5$ m²sec². The wind speed has a Weibull probability density function.

a) Calculate the scale factor, C

b) Calculate the shape factor, K

3.9 For a certain site, the average wind speed is 7.063 m/sec, at a standard height of 10 m; a wind energy conversion system with regular synchronous generator is installed. The wind energy conversion parameters are:

Cut-in wind speed: 4.5 m/sec
Rated wind speed: 10.8 m/sec
Furling wind speed: 22 m/sec
Hub height: 23 m
Rated generator output: 100 kW

a) Calculate the capacity factor of the WECS.

3.10 A wind farm has the installed capacity of 5 MW, the load factor is 30%. Calculate the amount of energy it produces in MWh in a year.

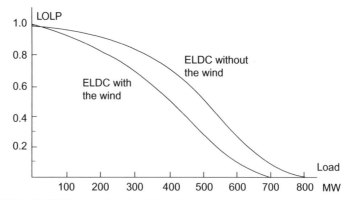

FIGURE 3.69 WECS equivalent load duration curves.

References

[1] http://people.bath.ac.uk/ft212/Website/persian%20windmill.gif; 2012 [accessed 10.08.12].

[2] Wind energy applications guide by American Wind Energy Association, http://media. photobucket.com/image/recent/NBBooks/WTGMajorComponents-1.jpg; 2012 [accessed 10.08.12].

[3] Southern Cross Industries, www.southx.co.za/photos.htm; 2010 [accessed 21.05.10].

[4] Synergy Power Corporation. Windturbines; 2010. www.synergypowercorp.com/ [accessed 21.05.10].

[5] Borowy B, Salameh Z. Methodology for the optimally sizing the combination of a battery bank and PV array in a Wind/PV hybrid system. IEEE Trans Energy Conversion 1996; 11(2):367–75.

[6] Community water supply in Niama District, Oujda, NE Morocco, www.bergey.com/ examples/niama.html; 2010 [accessed 21.05.10].

[7] Wind turbine tower on Xiao Qing Dao Island, www.nrel.gov/data/pix/searchpix.cgi? getrec=6585188&display_type=verbose.

[8] Metaefficien. The world's largest wind turbine, www.metaefficient.com/news/new-record-worlds-largest-wind-turbine-7-megawatts.html-277; 2010 [accessed 21.05.10].

[9] World Wind Energy Association, www.wwindea.org/; 2010 [accessed 21.05.10].

[10] NASA. U.S. Wind energy program for large horizontal-axis wind turbines, www.nasa. gov/vision/earth/technologies/wind_turbines.html; 2010 [accessed 21.05.10].

[11] Wikipedia. Wind turbine design; 2010. http://en.wikipedia.org/wiki/Wind_turbine_ design[accessed 21.05.10].

[12] Wikimedia Commons. Mod-5B Wind Turbine.jpq, http://commons.wikimedia.org/wiki/ File:Mod-5B_Wind_turbine.jpg; 2010 [accessed 21.05.10].

[13] IEEE Xplore Digital Library. The development of the 7.3 MW Mode-5A wind turbine generator system; 2010. http://ieeexplore.ieee.org/iel5/2201/4645927/04645943.pdf?arnumber [accessed 21.05.10].

[14] Salameh ZM, Safari I. Optimum windmill-site matching. IEEE Trans Energy Conversion 1992;7(4):669–76.

[15] Salameh Z, Borowy B, Amin A. Photovoltaic module-site matching based on capacity factor. IEEE Trans Energy Conversion 1995;10(2):326–32.

[16] Borowy BS, Salameh ZM. Optimum photovoltaic array size for a hybrid Wind/PV system. IEEE Trans Energy Conversion Sept. 1994;9(3):482–8.

[17] Salameh Z, Kazda L. Analysis of the steady state performance of the double output induction generator. IEEE Trans Energy Conversion 1986;EC-1:26–32.

[18] Borowy BS, Salameh ZM. Methodology for optimally sizing the combination of a battery bank and PV array in a Wind/PV hybrid system. In: IEEE power engineering summer meeting, Portland, Oregon; July 1995 95 SM 460-6-EC.

[19] Bagul A, Salameh ZM, Borowy BS. Sizing of PV and battery storage for a stand-alone hybrid wind-photovoltaic system, paper submitted and accepted for publication in Solar Energy.

[20] Anderson MB, Powles SJR. Wind energy conversion. Beccles, Suffolk: Waveney Print Services Ltd; 1986.

[21] Harry A. Energy conversion systems. John Wiley & Son, Inc; 1983.

[22] Telenet, http://users.telenet.be/b0y/content/gen_techin/Induction.Motor.cutaway.jpg; 2010 [accessed 21.05.10].

[23] Danish Wind Industry Association, Off-Shore wind farm, www.windpower.org.

[24] Integrated Publishing Inc., www.tpub.com/neets/book5/32NE0444.GIF; 2012 [accessed 10.08.12].

[25] American Windmill, www.awwasc.com/images/wind/app_5.jpg; 2012 [accessed 10.08.12].

[26] http://holytrinity.faithweb.com/p-on-Ecology/w01-aioliki.htm; [accessed 10.08.12].

[27] http://windeis.anl.gov/guide/maps/images/wherewind800.gif; [accessed 10.08.12].

[28] Wind Engineering & Consultancy, http://engineeringthewind.com/?page_id=9; 2012 [accessed 10.08.12].

[29] Wind Resource, Wind shear/Log-law, www.mstudioblackboard.tudelft.nl/duwind/Wind%20energy%20online%20reader/Static_pages/wind_shear.htm; 2012 [accessed 10.08.12].

[30] Department of Primary Industries, www.dpi.vic.gov.au/_data/assets/image/0009/34587/cross-section.jpg; 2012 [accessed 10.08.12].

[31] www.allatsea.net/southeast/blackford-windmill-boat/; 2012 [accessed 10.08.12].

[32] www.nrel.gov/data/pix/searchpix.php?getrec=10769&display_type=verbose; 2012 [accessed 10.08.12].

[33] www.highton.com/pages/portfpages/altampassportf.html; 2012 [accessed 10.08.12].

[34] Solaripedia, www.solaripedia.com/223/copyright.html; 2012 [accessed 10.08.12].

[35] Sciencedaily, www.sciencedaily.com/releases/2007/12/071212201424.htm; 2012 [accessed 10.08.12].

[36] Industrial Wind Action Group, www.windaction.org/pictures/13943; 2012 [accessed 10.08.12].

[37] http://universe-review.ca/I13.24-angles.jpg; 2012 [accessed 10.08.12].

[38] www.nasa.gov/vision/earth/technologies/wind_turbines.html; 2012 [accessed 10.08.12].

[39] http://us.ask.com/wiki/NASA_wind_turbines; 2012 [accessed 10.09.12]. photo: NASA/Glen Research Center. In 1975.

[40] Congress, www.telosnet.com/wind/govprog.html; 2012 [accessed 10.09.12].

[41] Amp Air Corporation, www.ampair.com/; 2012 [accessed 10.09.12].

[42] http://en.wikipedia.org/wiki/Great_GransdenNew; 2012 [accessed 10.09.12].

[43] New Energy Foundation, www.nef.or.jp/english/award2001/00_07.html; 2012 [accessed 10.09.12].

[44] www.bergey.com/niama-district-ne-morocco; 2012 [accessed 10.09.12].

[45] www.bergey.com/krasnoe-village-russia.

[46] http://article.wn.com/view/2012/07/24/Wind_farm_rises_on_Alaskan_island_1/; 2012 [accessed 10.09.12].

[47] http://optimal-power-solutions.com/blog/category/development/; 2012 [accessed 10.09.12].

[48] The Adventures of Marco Polie, http://marcopolie.blogspot.com/2010/12/ago.html; 2012 [accessed 10.09.12].

[49] http://en.wikipedia.org/wiki/File:Wb_deichh_drei_kuhs.jpg; 2012 [accessed 10.09.12].

[50] www.snowstudies.org/diarypages/04-05/remoteweatherstation.html; 2012 [accessed 10.09.12].

[51] http://divinecosmos.com/resources/divinecosmos/8.html; 2012 [accessed 10.09.12].

[52] www.telecomponentiromania.com/images/Turbine-Elicoidali-di-Savonius.PNG; 2012 [accessed 14.10.12].

[53] https://solarconduit.com/shop/wind/urban-green-energy-vawt-uge4kw.html; [accessed 14.10.12].

[54] http://multivu.prnewswire.com/mnr/wepower/39994/; [accessed 14.10.12].

[55] www.igreenspot.com/helix-wind-turbine-to-power-cell-phone-towers/; [accessed 14.10.12].

[56] www.interfacebus.com/Glossary-of-Terms-Rotor-Diagram.html; [accessed 14.10.12].

[57] www.mpoweruk.com/motorsac.htm; [accessed 14.10.12].

[58] www.windspireenergy.com/; [accessed 14.10.12].

[59] www.bizearch.com/trade/TWG_Series_Three_Phase_Brushless_AC_Generator_3080_
2938.htm; [accessed 14.10.12].

[60] www.ewh.ieee.org/soc/es/Nov1998/08/syncmach.htm; [accessed 14.10.12].

[61] https://sites.google.com/site/smabhyan/allaboutmotors; [accessed 14.10.12].

Energy Storage

4.1 BATTERY TECHNOLOGY

A battery, in concept, can be any device that stores energy for later use. A rock, pushed to the top of a hill, can be considered a kind of battery, since the energy used to push it up the hill (chemical energy, from muscles or combustion engines) is converted and stored as potential kinetic energy at the top of the hill. Later, that energy is released as kinetic and thermal energy when the rock rolls down the hill.

4.1.1 Introduction

Common use of the word, *battery*, however, is limited to an electrochemical device that converts chemical energy into electricity by use of a galvanic cell. A galvanic cell is a fairly simple device consisting of two electrodes (an anode and a cathode) and an electrolyte solution. Batteries consist of one or more galvanic cells.

The battery was developed in the late eighteenth century. The cause was championed by the work carried out by Luigi Galvani from 1780 to 1786. Through his experiments, Galvani observed that, when connected pieces of iron and brass were applied to frog's legs, they caused them to twitch. However, Galvani thought that the effect originated in the leg tissue. Nevertheless, Galvani had laid the cornerstone for further developments in "voltaic" electricity.

Most historians date the invention of batteries to about 1800 when experiments by the Italian professor Alessandro Volta (Figure 4.1), at the Pavia University, resulted in the generation of electrical current from chemical reactions between dissimilar metals. The original "voltaic pile" used zinc and silver disks and a separator consisting of a porous nonconducting material saturated with salt water. Volta described it as a "construction of an apparatus of unfailing charge, of perpetual power." When stacked, a voltage could be measured across each silver and zinc disk (Figure 4.2). The creation converted chemical energy into electrical energy. The more piles Volta assembled together, the bigger the jolt he could produce. He got his reward. Napoleon Bonaparte made Volta a Count and the unit of electrical measurement, the "Volt," was named in his honor. This is still the basis for most modern batteries.

Experiments with different combinations of metals and electrolytes continued over the next 60 years. Even though large and bulky variations of the voltaic pile

FIGURE 4.1 Count Alessandro Volta (1745–1827).

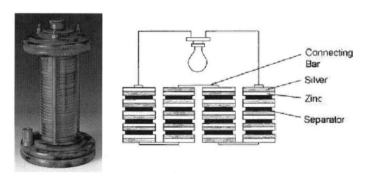

FIGURE 4.2 The voltaic pile.

provided the only practical source of electricity in the early nineteenth century, they were the original primary battery.

Johann Ritter first demonstrated the elements of a rechargeable battery in 1802, but rechargeable batteries remained a laboratory curiosity until the development, much later in the century, of practical steam-driven dynamos to recharge them.

By 1813, Sir Humphrey Davy constructed a large battery consisting of 2000 pairs of plates that occupied a surface area of 889 feet. He used this and other early batteries to gain a deeper understanding of the concepts of basic electricity. Davy's experiments helped explain how elementary substances, such as oxygen and hydrogen, combine through electrical attraction to form natural compounds such as water.

In the years that ensued, other means of producing electricity were invented, all of which involved the use of liquid electrodes. Those developed by Bunsen (1842) and Grove (1839) were among the most successful systems, and they were used for many years.

During the first half of the nineteenth century, experiments continued with a variety of electrochemical couples (combinations of positive and negative electrode materials and electrolytes). Finally, about 1860, the ancestors of today's primary and secondary batteries were developed.

By 1866, Georges Leclanche, a French engineer, patented a primary battery that was immediately successful. In the space of two years, 20,000 of his cells were being used in the telegraph system. It became popular because it was rugged, easy to manufacture, and had a good shelf life. Up until this time, one of the components of a battery was always a liquid. Leclanche substituted paste so that it was portable and would not leak. His original cell was assembled in a porous pot.

The positive electrode consisted of crushed manganese dioxide with a little carbon mixed in. The negative pole was a zinc rod. The cathode was packed into the pot, and a carbon rod was inserted to act as a currency collector. The anode or zinc rod and the pot were then immersed in an ammonium chloride solution. The liquid acted as the electrolyte, readily seeping through the porous cup and making contact with the cathode material. Leclanche's "wet" cell, as it was popularly referred to, became the forerunner to the world's first widely used battery—the carbon-zinc cell. The original design was improved to incorporate the electrolyte into a wet paste.

Leclanche's invention, which was quite heavy and prone to breakage, was steadily improved over the years. The idea of encapsulating both the negative electrode and porous pot into a zinc cup was first patented by J. A. Thereabout in 1881. But, it was Carl Gassner of Mainz who is credited as constructing the first commercially successful carbon-zinc "dry" cell. Variations followed. By 1889, there were at least six well-known dry batteries in circulation. Later battery manufacturing produced smaller, lighter ones, and the application of the tungsten filament in 1909 created the impetus to develop batteries for use in flashlights. The carbon-zinc dry cell is still the mainstay of the primary battery market.

The production of batteries was greatly increased during World War I as a means of powering flashlights and field radios. Other milestones in battery production include widespread radio broadcasting, which brought battery-operated wireless into many homes. But, it was during the interwar years that battery performance was greatly enhanced. This was achieved through better selection of materials and manufacturing method.

Shortly before World War II, Samuel Ruben invented the mercury primary cell. At first, mercury cells were expensive, and due to their small size, they were not used in many applications. With the invention of the transistor during the 1950s, mercury cells found their way into hearing aids and transistor radios.

The alkaline manganese cell was further refined in the 1950s. At the same time, small cameras with built-in flash units, which required high power in a small package, were developed. Alkaline cells worked in this and other new consumer applications so well that they gained tremendous popularity. They remain one of the largest source of revenue for portable battery sales to this day.

Secondary (or rechargeable) batteries date back to 1860 when the French physicist, Raymond Gaston Planté, invented the lead-acid battery. His cell used two thin lead plates separated by rubber sheets. He rolled the combination up and immersed it in a dilute sulfuric-acid solution. Initial capacity was extremely limited since the positive plate had little active material available for reaction. About 1881, Faure and others developed batteries using a paste of lead oxides for the positive plate's active materials. This allowed much quicker formation and better plate efficiency than the solid Planté plate.

Although the rudiments of the flooded lead-acid battery date back to the 1880s, there has been a continuing stream of improvements in the materials of construction and the manufacturing and formation processes. The lead-acid battery is the most popular battery in use today and can be found in everything from automobiles to wheelchairs.

Around 1890 Thomas Edison began perfecting batteries as a way of powering his new invention, the phonograph. It took him nearly 20 years to perfect the alkaline-storage battery, but by 1909 he was selling batteries to power submarines and electric vehicles. The nickel-iron cell is unique because it has a long life and can be used for heavy industrial applications. This battery had an alkali paste (an organic material) instead of acid. Along the way, Edison perfected different batteries for many uses. His batteries even powered the first self-starting Model T Ford in 1912.

In 1899, Waldmar Jungner invented the nickel-cadmium rechargeable battery. It was expensive compared to other batteries, and its use was limited. In the 1930s, new electrodes were developed, and in the 1940s a sealed nickel-cadmium battery, which recombines internal gases produced during charge, was perfected. Steady improvements have been made every decade since.

Because many of the problems with flooded lead-acid batteries involved electrolyte leakage, numerous attempts have been made to eliminate free acid in them. German researchers developed the gelled-electrolyte lead-acid battery in the early 1960s; it was a major improvement. Working from a different approach, Gates Energy Products developed a sealed-lead battery, which represents the state-of-the-art today.

Development of the nickel–metal hydride (niMH) rechargeable cell began in the 1970s, but it was a long time before hydride alloys performed well enough to begin production. Since the late 1980s, the performance of niMH cells has steadily improved, and there may still be room for further performance advances.

The latest developments in batteries, both primary and rechargeable, have centered on the use of lithium. Lithium is the lightest of all metals, has the greatest electrochemical potential, and provides the most energy. Lithium primary batteries were popularized during the 1970s and 1980s. They have replaced the alkaline cell in most photo applications and are better suited to military and scientific products than any other type.

Attempts to make lithium rechargeable batteries go back to the 1980s. Problems with safety prevented the commercial use of the technology at that time. Finally, rechargeable cells that use lithium metal were abandoned. Research shifted to the use of lithium ions found in chemicals such as lithium-cobalt dioxide. Since then lithium-ion batteries have become the most popular choice for use in high-tech applications such as cellular phones and laptop computers.

Batteries have now become an essential part of everyday life. They are the power source for millions of consumer, business, medical, military and industrial appliances worldwide. This demand is growing.

4.1.2 **Battery construction and principles of operation**

Battery action takes place in the cell, the basic battery building block that transforms chemical into electrical energy. A cell contains the two active materials or electrodes and the solution electrolyte that provides the conductive environment between them. There are two kinds of batteries: primary and secondary. In the primary type, the chemical reaction eats away one of the electrodes (usually the negative), and the cell must be discarded or the electrodes replaced; however, in the secondary type, the chemical reaction is reversible, and the active electrodes can be restored to their original condition by recharging the cell. A battery can consist of only one cell the same as the battery that operates the flashlight, or several cells in a common container such as the secondary battery that powers an electric vehicle (EV).

Active materials

The active materials are defined as electrochemical couples. This means that one of the active materials, the positive pole or anode, is electron-deficient; the other active materials, the negative pole or the cathode, is electron-rich. The active materials could be solid as in the lead-acid battery, liquid as in the sodium–sulfur battery, or gaseous as in the zinc-air battery. When the load is connected across the battery, the voltage produces an external current flow from positive to negative corresponding to its internal electron flow from the negative pole to the positive pole.

The observed voltage is the sum of what is happening at the anode and the cathode; however, to make an ideal battery, the active materials have to be chosen so that they give the greatest oxidation potential at the anode coupled with the materials that give the greatest reduction potential at the cathode—that is, coupling the best reducing materials (lithium) with the best oxidizing materials (fluorine).

Electrolyte

The electrolyte provides the path for the migration of the electrons between the electrodes and, in some cells, also participates in the chemical reaction. The electrolyte is usually a liquid phase; however, it could be jelly or paste form. In terms of chemistry, a battery is electrodes and electrolytes operating in a cell or container in accordance with certain chemical reactions. A battery consists of one or more electrochemical cells. Although the terms *battery* and *cell* are often used interchangeably, cells are the building blocks on which batteries are constructed. Batteries consist of one or more cells that are electrically connected.

Cell

A cell normally consists of the four principal components shown in Figure 4.3:

- A positive electrode that receives electrons from the external circuit when the cell is discharged
- A negative electrode that donates electrons to the external circuit as the cell discharges
- An electrolyte that provides a mechanism for charge to flow between positive and negative electrodes
- A separator that electrically isolates the positive and negative electrodes

In some designs, physical distance between the electrodes provides the electrical isolation and the separator is not needed. In addition to the critical elements listed here, cells intended for commercial batteries normally require a variety of packaging and current-collection apparatus to be complete.

How a cell works. When a battery or cell is inserted into a circuit, it completes a loop, which allows charge to flow uniformly around the circuit. In the external part of the circuit, the charge flow is electrons resulting in electrical current. Within the cell, the charge flows in the form of ions that are transported from one electrode to the other. As mentioned before, the positive electrode receives electrons from the external circuit on discharge. These electrons then react with the active materials of the positive electrode in "reduction" reactions that continue the flow of charge through the electrolyte to the negative electrode. At the negative electrode, "oxidation"

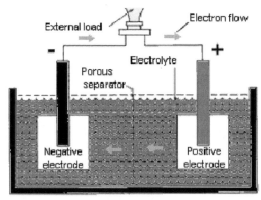

FIGURE 4.3 Principle of operation of a battery cell.

reactions between the active materials of the negative electrode and the charge flowing through the electrolyte result in surplus electrons that can be donated to the external circuit.

It is important to remember that the system is closed. For every electron generated in an oxidation reaction at the negative electrode, there is an electron consumed in a reduction reaction at the positive. As the process continues, the active materials become depleted and the reactions slow down until the battery is no longer capable of supplying electrons. At this point, the battery is discharged.

Battery characteristics

Several terms are used when discussing the characteristics of batteries. This is important in determining which battery should be used for a particular application. The following are short description of these characteristics.

Voltage, V: The voltage is a unit of measurement of electrical potential difference between any two points. It is also known as the *electromotive force*. The electrical potential between the anode and the cathode in the batteries is called the *battery voltage*. Different battery cell generate different voltages, the higher the better.

Capacity, C: It is how many amperes (amp) a battery can provide. It is measured in Ah. The battery capacity depends on:

- Area and the physical size of the plates in contact with the electrolyte.
- Weight and the amount of material in the plates.
- Number of plates and the type of separator between them.
- Quantity and specific gravity of the electrolyte; the specific gravity of a material is the density of the material divided by the density of the water; the hydrometer is used to measure the specific gravity.
- The age of the battery; the older the battery the less capacity it has.
- Cell conditions: sulfation, sediments reduce the cell voltage.
- Temperature: the higher the temperature the higher the capacity.
- Discharge rate: the higher the discharge rate, the lower the capacity.

State of charge, SC: This is how much energy the battery has—measured in %. Battery voltage is the best measure of the state of charge, the higher the voltage the higher the state of charge.

Rate of charge/discharge current, C/T: It is a measure of how fast you charge or discharge a battery. Also, the higher the rate of charge, the higher the battery voltage at the end of the charging process. The rate of charge or discharge is calculated as the ratio of the capacity of the battery over the time it takes to charge/discharge it. The higher the rate of discharge, the lower the voltage at the end of the discharging process.

Cutoff voltage, V_{co}: The voltage below which, when a battery gets discharged to, it gets damaged. For a lead-acid battery, it is 10.5 V.

Depth of discharge, DOD: The percentage of the battery energy taken during the discharging process before the battery can be damaged; it is measured in %.

Lead-acid batteries can be discharged to 80%. NiCd batteries can be discharged to 100%.

Resistance and impedance: The property of the battery material to resist the flow of current is called the battery resistance. The battery has a DC resistance and impedance. The battery resistance is usually around 0.12 ohms. It has a DC and an AC component.

Efficiency: The efficiency is ratio of the energy out to the energy in. Lead-acid batteries are 80 to 85% efficient.

Self-discharge, SD: The battery discharges itself, while it is on the shelf without being used. Some batteries have higher self-discharge rates than others. niMH batteries have a very high rate of self-discharge.

Specific energy: This is the total amount of energy in watt-hours or kW-hours the battery can store per kilogram of its mass for a specified rate of discharge. An important factor in determining the range of an electric vehicle powered by batteries.

Specific power: This the gravimetric power density and it is expressed as W/kg.

Energy density Refers to the amount of energy a battery has in relation to its size. Energy density is the total amount of energy (in Wh/liter) a battery can store per liter of its volume for a specified rate of discharge. Batteries that have high energy density are smaller.

Power density This is the volumetric power density of a battery and expressed as W/liter.

Peak specific power: Sustained peak power for 30 sec discharge without reaching 80% capacity. This is important for the acceleration of electric vehicles. Peak specific power is the maximum number of watts per kilogram (W/kg) a battery delivers at a specified depth of discharge. Specific power is at its highest when the battery is fully charged. As the battery is discharged, the specific power decreases and acceleration decreases also. Specific power is usually measured at 80% depth of discharge.

Cycle life: The total number of times a battery can be discharged and charged during its life. When the battery can no longer hold a charge of more than 80%, its cycle life is considered finished.

4.1.3 Applications

Batteries come in all sizes and shapes to fit a diversity of applications. However, there are three major parameters that work to determine the suitability of battery types for an application. The first major concern is whether the battery is used in high-rate or low-rate service. Some batteries are much better suited to handling a long-term drain at a low rate than they are a high-rate load for a short interval. Examples of low-drain situations include memory backup for electronic circuits and clocks or watches. High-rate loads include most cordless appliances and engine starter.

The second concern pertains to rechargeable batteries. It is the relative amount of time the battery spends being charged versus the time it spends on discharge. Some batteries are used in "float" applications where they spend most of their time on charge with only rare discharges. Most power backup applications fall into this category. "Cyclic" applications are those where the battery is used regularly and it gets relatively little time to recharge between uses. Most battery-powered portable equipment falls into this category.

The last major concern regarding battery applications is the environment in which it is used, specifically the temperatures at which the battery is required to operate. Batteries and people like approximately the same temperature range. If the temperature gets too warm, the chemical reactions within the battery are accelerated and its life may be shortened. If the battery gets too cold, the chemical reactions are slowed down, reducing battery output.

4.1.4 Types of batteries used for energy storage

There are many different types of batteries, with new formulations being developed all the time. For energy storage the following subsections describe the types of batteries that are used.

Lead acid

Lead-acid is one of the oldest and most developed battery technologies. They remain popular because they can produce high or low currents over a wide range of temperatures, they have a good shelf life and life cycles, and they are relatively inexpensive to manufacture and purchase. Lead-acid batteries are usually rechargeable. It is a low-cost and popular storage choice for power quality, uninterruptible power supply (UPS), and some spinning reserve applications. Its application for energy management, however, has been very limited due to its short cycle life. The amount of energy (kWh) that a lead-acid battery can deliver is not fixed and depends on its rate of discharge.

The most widely known uses of lead-acid batteries are as automobile batteries. Rechargeable ones have been available since the 1950s and have become the most widely used type of battery in the world—more than 20 times the use rate of its nearest rivals. In fact, battery manufacturing is the single largest use for lead in the world.

Lead-acid batteries come in all manner of shapes and sizes, from household batteries to large batteries for use in submarines (Figure 4.4). The most noticeable shortcomings of them are their relatively heavy weight and their falling voltage during discharge. Lead-acid batteries, nevertheless, have been used in a few commercial and large-scale energy management applications.

The largest one is a 40 MWh system in Chino, California, built in 1988 (Figure 4.5). There are several manufacturers who have developed lead-acid storage systems that are larger than 1 MWh. This battery is an ambient temperature, aqueous electrolyte battery. A cousin to this battery is the deep-cycle lead-acid battery, now widely used in golf carts, forklifts, and some of the EVs available today. The first electric cars built used this technology.

FIGURE 4.4 Typical lead-acid batteries (Hawker, Sonnenshine, Delphy, and Electrosources).

FIGURE 4.5 Inside views of Chino's 10-MW, 40-MWh lead-acid battery storage system in California.

Lead-acid batteries are inexpensive, readily available, and are highly recyclable, using the elaborate recycling system already in place. Although inexpensive, it is very heavy, with a limited usable energy by weight (specific energy). The battery's low-specific energy and poor energy density make for a very large and heavy battery pack, which cannot power a vehicle as far as an equivalent gas-powered vehicle. Lead-acid batteries should not be discharged below 80% of their rated capacity or DOD. Exceeding the 80% DOD shortens the life of the battery. Research continues to try to improve these batteries.

A lead-acid no aqueous (gelled lead-acid) battery, uses an electrolyte paste instead of a liquid. These batteries do not have to be mounted in an upright position.

There is no electrolyte to spill in an accident. No aqueous lead-acid batteries typically do not have as high a life cycle and are more expensive than flooded deep-cycle lead-acid batteries.

There are many benefits to sealed lead-acid (SLA) battery. SLA isn't affected by the so called "memory effect" that nickel batteries suffer from, and SLA has the best charge retention of any rechargeable battery. With SLA, it is fairly easy to determine the remaining capacity based on open circuit voltage (OCV); that is only about 20% accurate though.

There are some down sides to SLA. Besides the size and weight disadvantage, sealed lead-acid batteries won't take a fast charge well and do not perform well at low temperatures. SLA loses its charge quickly if stored at high temperatures. The exceptions are "pure-lead" batteries, which are capable of fast charging and perform well at lower temperatures. Pure-lead batteries also have much better charge retention and recovery characteristics after long periods of storage (one or two years).

SLA batteries prefer a constant voltage charge. VRLA batteries can take a charge well from 7 hours at one-fourth the battery's capacity in input current, C/4, to 14 hours at C/10 or one-tenth the battery's rated capacity in input current. At these charge rates, 2.45 V per cell at 20 °C is best. C/300 is a good trickle or float charge current with 2.35/cell at 20 °C. The voltage of lead-acid cell goes down during discharge; this process is affected by discharge rates, the higher the discharge rate the lower the voltage at a certain depth of discharge (Figure 4.6). The lead-acid battery voltage is also affected by temperature; the higher the temperature during discharge, the higher the voltage at a certain discharge rate (Figure 4.7).

Temperature has a direct affect on the performance of a lead-acid battery. The concentration of sulfuric acid inside it increases and decreases with temperature. The most efficient temperature that battery manufacturers recommend is 78 °F. Because the temperature factor is important in colder climates, insulated battery

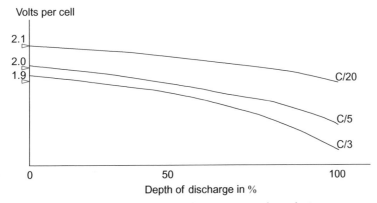

FIGURE 4.6 Voltage per cell versus depth of discharge for various discharge rates.

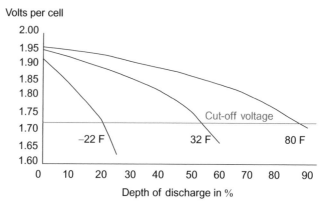

FIGURE 4.7 Voltage cell versus depth of discharge for various temperatures.

boxes, or thermal management systems, are highly recommended. During discharge, the battery voltage rises higher with higher discharge rates as seen in Figure 4.8. The battery voltage is a good indicator of its state of charge, although this is also affected by the rate of charge. The higher the rate of charge, the higher the voltage at a certain state of charge (Figure 4.9).

Lead-acid batteries are the least expensive ones available. Their name, as is the case with all batteries, comes from their construction. The positive electrode (anode) is made up of lead dioxide (PbO_2). The negative electrode is made up of Pb. The electrolyte is a dilute solution of sulfuric acid (H_2SO_4). The chemical reactions that take place on the anode and the cathode during charging and discharging are the following:

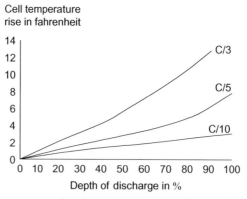

FIGURE 4.8 Battery temperature rise during discharge for different discharge rates.

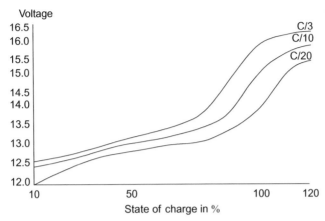

FIGURE 4.9 Battery voltage under charge versus state of charge for different charge rates.

Discharge

Cathode: $Pb + SO_4^{(2-)} \Rightarrow PbSO_4 + 2e^-$

Anode: $PbO_2 + 2e^- + SO_4^{(2-)} + 4H^+ \Rightarrow PbSO_4 + 2H_2O$

Charge

Cathode: $PbSO_4 + 2e^- \Rightarrow Pb + SO_4^{(2-)}$

Anode: $PbSO_4 + 2H_2O - 2e^- \Rightarrow PbO_2 + SO_4^{(2-)} + 4H^+$

As we can clearly see, the reactions are reversed when going from charging to discharging.

Lead-acid batteries consist of six cells connected in series. Each cell has a voltage of 2 V, thus bringing the total voltage of the battery to 12 V. In reality, however, the voltage ranges from about 10.5 V when fully discharged to 15 V when fully charged. Additional properties of lead-acid batteries and their applications include:

Cell voltage: 2.0 V (nominal)

Capacity: 500–100 mAh or more

Energy by weight: 30 Wh/kg

Energy by volume: Wh/cm^3

Cycle life: 200–500 cycles

Self discharge: 5%/month

Temperature range: $-20\,°C$ to $+60\,°C$

Preferred charge methods: constant voltage C/10 to C/4

Applications: communication equipment, office equipment, security systems, large and small power tools, toys, UPS systems, backup for lighting equipment (used with solar cells), CATV, CVCF, emergency lights, engine starting (portable generator), PBX, base stations, audio equipment, VCRs

Nickel cadmium

The nickel–cadmium (NiCd) battery is a well-established chemistry. Although newer chemistries have taken the spotlight in recent years, NiCd remains as viable today as it was years ago. Some sources sight more than 50% of the world's rechargeable batteries for portable applications are NiCd. The basic galvanic cell in a NiCd battery contains a cadmium anode, a nickel–hydroxide cathode, and an alkaline electrolyte. Batteries made from NiCd cells offer high currents at relatively constant voltage, and they are tolerant of physical abuse.

NiCd batteries have several advantages over other chemistries in use today. NiCds can take a fast charge, some in as little as 15 minutes and can take a charge at lower temperatures. Maintained properly, NiCds are capable of thousands of cycles. They can sustain heavy loads. Pricewise, NiCd is a bargain compared to some newer chemical formulas.

Perhaps most important, though, is the wide variety of cells and construction methods available with NiCd; no other chemistry comes close in terms of selection. Most NiCd batteries used for portable application are either spiral-wound cylindrical cells or button cells. There are larger plate NiCds—wet cells used in aircraft and long-term stationary applications.

Several kinds of electrodes are used in NiCd cells. Sintered, plastic bonded, and foam are the main types. By combining different types of electrodes in a cell, manufacturers are able build ones that have unique advantages over other types. For example: NiCd cells may be tailored for superior performance at high temperature, high discharge, fast charge, long-term float or trickle charge, or may be made to deliver extra capacity and be economically priced (Figure 4.10).

The reputation of the NiCd has suffered over the years from the so-called "memory effect"; the term implies that the battery "remembers" how much energy was required on previous discharges. Simply put, memory effect is a loss of NiCd cell performance after relatively few cycles. This is usually caused by *crystalline formation* on the electrodes of the cell. When the crystals grow, they reduce the surface area of the electrodes, which leads to voltage depression and loss of performance.

The biggest cause of memory effect is overcharge due to the battery being left on charge for an indefinite period of time or being charged before necessary. The best remedy is regular use or complete discharge to 1.0 V/cell. Both cyclic and standby batteries should be fully discharged once a month to maintain optimum performance. If memory effect becomes really bad because of improper maintenance, there are a host of excellent battery analyzer/reconditioners that should be able to repair the damage. If too much time has passed, however, even fancy equipment won't bring your NiCd back.

Under normal conditions NiCd batteries prefer a constant current charge. Several cycles may be necessary before it reaches its fully rated capacity. Standard NiCd batteries take a 14-hour charge with an input current of 1/10 of the battery's capacity or a C/10 rate. Special NiCd batteries will take a charge in as little as an hour or less with proper termination. The best termination for NiCd is when the charge

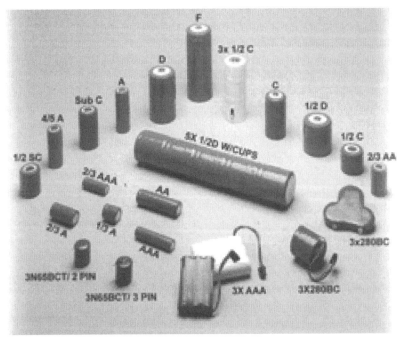

FIGURE 4.10 Examples of NiCd batteries.

stops after the battery reaches peak voltage and just starts to drop in voltage as it goes into overcharge. NiCd batteries should be charged in temperatures between $5\,^\circ C$ to $+45\,^\circ C$.

Unfortunately, nickel–cadmium technology is relatively expensive. Cadmium is an expensive metal and is toxic. Recent regulations limiting the disposal of waste cadmium (from cell manufacturing or from disposal of used batteries) have contributed to the higher costs of making and using these batteries.

The second type of battery available for use in EVs and enegy storage is nickel–cadmium one. The cathode of these batteries is made up of cadmium, Cd. The anode is made up of nickel–oxide hydroxide, NiOOH. The electrolyte is made up of potassium hydroxide, KOH. The following shows the chemical reaction in a nickel cadmium cell:

Discharge

Anode: $2NiOOH + 2H^+ + 2e^- => 2Ni\,(OH)_2$

Cathode: $Cd + 2OH^{(2-)} => 2e^- + Cd\,(OH)_2$

Charge

Anode: $2Ni\,(OH)_2 - 2e^- => 2NiOOH + 2H^+$

Cathode: $Cd\,(OH)_2 + 2e^- => Cd + 2OH^{(2-)}$

One important disadvantage of the NiCad batteries is the fact that they display memory effects. Some other particular characteristics of NiCad batteries are the following:

Cell voltage: 1.25 V (nominal)
Capacity: lama to 20,000 mAh
Energy by weight: 40–60 Wh/kg
Energy by volume: Wh/cm^3
Discharge characteristics: 100% DOD
Self-discharge: 20%/month

Self-discharge characteristics of Energizer nickel–cadmium cells are shown in Figure 4.11. The characteristics are shown as a decline in percent of rated capacity available. Self-discharge is increased by elevated temperatures. Batteries are not harmed even if not used for long periods of time:

Temperature range: $-40\,°C$ to $+60\,°C$
Preferred charge methods: constant current
Sizes: widest selection of cylindrical and button sizes

The charging of the battery is an exothermic process, while the discharging is endothermic. They display dendrite growth. This means that whiskers grow from the electrodes. To get rid of them, we need to charge the battery at a high rate of C/0.02.

Longevity: 10 years (2000 cycles)
Cost: expensive
Resistance: very low internal resistance
Cutoff voltage: $\sim$1 V per cell
Applications: Cameras, data terminals, FAX and POS memory, hobby remote controls, Notebook PCs, portable phones, transceivers, portable printers, portable TVs, CD and tape players, security lights, power tools, vacuum cleaners, shavers, memory back-up, security systems, emergency systems, office equipment, toys, video camcorders

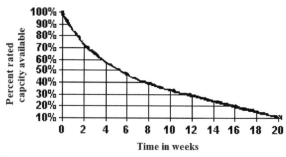

FIGURE 4.11 NiCd self-discharge rate.

Nickel cadmium batteries have certain advantages when compared to lead-acid batteries. They have a longer life, better low-temperature performance, lower internal impedance, and a higher energy density. However, there are two reasons that make NiCd batteries less appealing, and (1) high cost and (2) environment-related cadmium issues.

Sodium sulfur battery

A sodium sulfur (NaS) battery consists of liquid (molten) sulfur at the positive electrode and liquid (molten) sodium at the negative electrode as active materials separated by a solid beta alumina ceramic electrolyte. The electrolyte allows only the positive sodium ions to go through it and combine with the sulfur to form sodium polysulfides.

NaS batteries have a lot of good points: a high energy-storage capacity (45 Wh/kg +), do not discharge like the NiCd batteries, and are fairly inexpensive. They have a few very bad points to match that: NaS batteries have to run at temperatures where the sodium and sulfur are liquids—between 290 and 390 °C. In addition, there are serious safety considerations with carrying a load of highly reactive molten sodium in cars if it is used in EVs. Sodium explodes and burns when exposed to water and can cause very nasty burns. Typically, the sodium and sulfur are separated by a ceramic electrolyte; if this ceramic should break, the reaction between the parts of the cell generates a lot of heat. In addition, they can only be recharged $\sim$500 times before they need to be replaced.

Their cell voltage $= 2.076$ V, and the reactions of a NaS battery are very simple:

$$2Na = 2Na^+ + 2e^-$$

$$4S + 2e^- = S_4^{(2-)}$$

$$2Na + 4S = Na_2S_4$$

During discharge, as positive Na^+ ions flow through the electrolyte and electrons flow in the external circuit of the battery, about 2 V is produced. This process is reversible because charging causes sodium polysulfide to release the positive sodium ions back through the electrolyte to recombine as elemental sodium. The battery is kept at about 300 °C to allow this process.

Sodium sulfur battery cells are efficient ($\sim$89%) and have a pulse power capability more than six times their continuous rating (for 30 sec). This attribute enables the NaS battery to be economically used in combined power-quality and peak-shaving applications. NaS battery technology (Figure 4.12) has been demonstrated at more than 30 sites in Japan, totaling more than 20 MW with stored energy suitable for 8 hours daily peak shaving (Figure 4.13). The largest NaS installation is a 6 MW, 8-h unit for Tokyo Electric Power Company. Combined power-quality and peak-shaving applications in the US market are under evaluation. Commercial production of the basic building block—the NaS 50-kW, 360-kWh module—was targeted for early 2003.

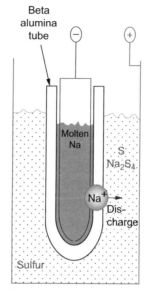

FIGURE 4.12 NaS module from NGK Technologies (Japan).

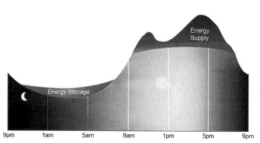

FIGURE 4.13 Peak shaving.

The Ford Motor Company uses sodium sulfur batteries in their Ecostar electric car, a converted delivery minivan that is currently sold in Europe. NaS batteries are only available to EV manufacturers. This battery is a high-temperature one with the electrolyte operating at temperatures of 572 °F (300 °C). The sodium component of this battery explodes on contact with water, which raises certain safety concerns. The materials of the battery must be capable of withstanding the high internal temperatures they create, as well as freezing and thawing cycles. The NaS battery has a very high specific energy: 50 Wh/lb (110 Wh/kg).

Vanadium redox battery

The vanadium redox battery (VRB) stores energy by employing couples, V^{2+}/V^{3+} in the negative and V^{4+}/V^{5+} in the positive half-cells (Figure 4.14). These are stored in mild sulfuric acid solutions (electrolytes). During the charge/discharge cycles, H^+ ions are exchanged between the two electrolyte tanks through the hydrogen-ion permeable polymer membrane. The cell voltage is 1.4 to 1.6 V. The net efficiency of this battery can be as high as 85%. Like other flow batteries, the power and energy ratings of the VRB are independent of each other.

Do not think of the VRB as a battery to power your watch, cell phone, or flashlight (torch). It is not intended to replace the 12 V lead-acid battery that supports the electrical system in your car. The VRB will be used for emergency standby power in the telecommunications industry, as the backup unit in UPS systems, a load-leveling device in a power station, the energy-storage device in a solar- or wind-generated remote power system, or to power the electric car. The VRB is a reduction–oxidation (redox) flow battery that employs an electrolyte solution where energy is stored and a cell stack where the energy conversion occurs.

The VRB essentially is comprised of three basic components: a negative electrolyte tank with a pump, a positive electrolyte tank with a pump, and a series of membranes or electrochemical cells ("the cell stack") through which the vanadium electrolyte flows (or is pumped). Each cell in the cell stack is an electrically conductive carbon impregnated polymer sheet to which graphite felt is heat bonded. The electrodes are inert and unlike conventional lead-acid batteries do not participate in the electrochemical reactions. Vanadium, an abundant and stable metal, is held in two ionic forms in a diluted sulfuric-acid electrolyte solution. It is the vanadium pentoxide resulting from this process that effectively stores the energy.

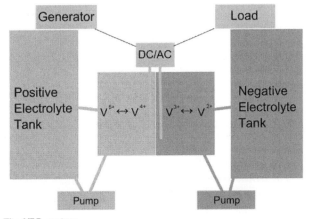

FIGURE 4.14 The VRB system.

The two electrolyte solutions circulate from the storage tanks, through the cell stack and back to the tanks. Energy is released when charged electrolyte flows through the cell stack causing electron transfer between the different forms of vanadium ions across the separating membrane to flow into an external load. Running current into the cell stack reverses the process and recharges the electrolyte solution, which can be reused to release energy at any time. Charging the battery or *storing* energy is possible from a number of different sources such as wind, solar, hydro, excess power generation, or mains electricity.

Deployment status VRB was pioneered in the Australian University of New South Wales (UNSW) in the early 1980s. The Australian Pinnacle VRB bought the basic patents in 1998 and licensed them to Sumitomo Electric Industries (SEI) and Vanteck. The vanadium redox battery storages up to 500 kW, 10 hrs (5 MWh) and have been installed in Japan by SEI. VRBs have also been applied for power-quality applications (3 MW, 1.5 sec., SEI). The first large commercial VRB installation outside Japan was done at ESKOM in South Africa by Vanteck (250 kW, 2 hrs).

Polysulfide bromide battery

Polysulfide bromide battery (PSB) is a regenerative fuel-cell technology that provides a reversible electrochemical reaction between two salt solution electrolytes—that is, sodium bromide and sodium polysulfide (Figure 4.15). Like other flow batteries, the power and energy ratings of Regenesys are independent of each other.

PSB electrolytes are brought close together in the battery cells where they are separated by a polymer membrane that only allows positive sodium ions to go through, producing about 1.5 V across the membrane. Cells are electrically

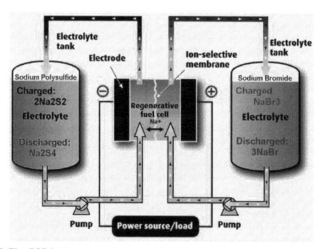

FIGURE 4.15 The PSB battery system.

With permission from Springer. James A. Kent Handbook of Industrial Chemistry and Biotechnology 12th ed. 201210.1007/978-1-4614-4259-2_38 © Springer Science+Business Media New York 2012. Renewable and Sustainable Energy Reviews, Volume 12, Issue 5, June 2008. The article title was "Energy storage systems —Characteristics and comparisons" by H. Ibrahima, A. Ilincaa, J. Perron.

connected in series and parallel to obtain the desired voltage and current levels. The net efficiency of this battery is about 75%, and it works at room temperature. It has been verified in the laboratory and demonstrated at multi-kW scale in the United Kingdom.

Chemical reaction

The polysulfide bromide battery will store or release electrical energy using a reversible electrochemical reaction between the two salt solutions or electrolytes. When the PSB battery is discharged, the electrolytes in the system are concentrated solutions of sodium bromide and sodium polysulfide. Conversely, the PSB battery is charged when the bromide ions are oxidized to bromine and complexed as tribromide ions and sulfur present in the polysulfide anion is converted to sulfide. The general equation for this reaction is:

$$3NaBr \rightarrow NaBr_3 + 2Na^+ + 2e^-$$

$$Na_2S_4 + 2Na^+ + 2e^- \rightarrow 2Na_2S_2$$

Regenesys Technologies is building a 120 MWh, 15 MW energy-storage plant at Innogy's Little Barford Power Station in the United Kingdom. The US Tennessee Valley Authority (TVA) is also planning to build a 12 MW, 120 MWh unit in Mississippi.

Zinc–bromine battery

The zinc–bromine battery (ZBB) consists of a negative zinc electrode and a positive bromine electrode. The electrodes are separated by a microporous separator. A solution, which is mostly aqueous with zinc and bromide salts dissolved into it, is then pumped into the cells from an anolyte reservoir and a catholyte reservoir. The chemical reactions take place in the reservoir (Figure 4.16).

The battery operates at an ambient temperature and when it is charging, metallic zinc plates the negative electrode. A major difference that distances the ZBB from lead and acid batteries is that the electrolytes do not take part in the reaction, so the electrode material does not deteriorate. When the battery has completely discharged, all of the zinc that has plated the negative electrodes has dissolved into the electrolyte. Thus, this battery's performance does degrade like most rechargeable batteries. Since it does not lose performance it has a wide spectrum of possible applications.

Some of the applications include electric cars, lawn mowers, scooters, motorized wheel chairs, and telecommunications supplies. Zinc–bromine batteries contain 2 to 3 times the energy density with respect to present lead-acid batteries. The battery has 75 to 85 Wh/kg. The measured potential difference is around 1.8 V per cell. When the battery is completely discharged, it can be left indefinitely. Today, ZBBs are capable of energy storage for 2 to 10 hours.

This advanced battery technology is still considered to be in its early stages of commercialization. The prototype systems range in size from 50 to 400 kWh. The projected system costs are expected to be under $400/kWh. The efficiency for ZBBs is 70%. The batteries offer six times the energy-storage capacity of lead-acid

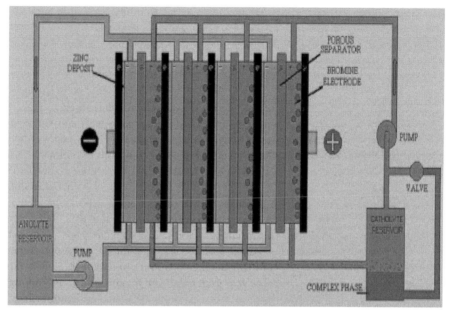

FIGURE 4.16 The zinc–bromide battery.

batteries. If a component fails, the battery can be refurbished; it doesn't have to be thrown away like other batteries.

In Australia, ZBB Energy Corporation and Australian Inland Energy & Water (AIEW) have received a $265,000 grant to field test a 500 kWh zinc–bromine battery to demonstrate how useful it could be in renewable energy applications. AIEW is hoping that this battery will eliminate the need for grid- of diesel-generator backup for renewable energy applications. The ZBB will be charged off of photovoltaic cells and from wind energy. Other things, which will be demonstrated, are the load-leveling capabilities of the battery.

The zinc–bromine battery designed by ZBB Energy for renewable energy consists of three battery stacks. Each stack has 60 cells in series. This configuration supplies 108 V. In most renewable energy applications, the battery will operate through power-conditioning equipment. An inverter is required if AC output is needed. The discharge capacity of the ZBB is 50 kWh. The battery can be charged at different rates up to 225 amps. This means that when the battery is in a fully discharged state it can be fully charged in a little as 3 hours. It can be continuously discharged at a rate of 300 amps. The battery comes equipped with a fully automated control and safety system.

The economic impact of the use of ZBBs is quite significant. If they were used at electric utilities, it is estimated $57 billion on the supply side could be saved. The use of zinc–bromine batteries for load leveling would be the equivalent of saving millions of barrels of oil a year. At higher production levels, the cost of ZBBs are priced competitively with lead-acid batteries but have a longer cycle life and are completely recyclable.

ZBB Energy is in a joint venture with China, where energy needs are dramatically increasing and the economy is growing. The Chinese's energy use has, on average, grown 8% over the last 20 years. This great increase in energy demand has created pollution problems with 13% of China's arable land affected by it. The priority there is to increase renewable energy sources. Under the joint venture agreement, ZBB's zinc–bromine batteries are being sold in China for energy storage.

When the battery is charging the zinc ion goes to the negative electrode and the bromine ion goes to the positive electrode. When the ZBB discharges the zinc ion flows to the positive electrode and the bromine ion flows to the negative electrode. Figure 4.17 shows the process of charging and discharging of the zinc–bromine batteries.

The two electrode cells are divided by a membrane, which could be the micro-porous membrane or an ion-exchange membrane to prevent the bromine from going to the positive electrode, where it would react with the zinc and discharge. In addition to the membrane, agents are added to the positive electrolyte to prevent the bromine from reaching the negative electrode. The complexing agents also help keep the pressure of the bromine down. The agents make the bromine from a red oily liquid. A ZBB can be optimized for power or energy delivery. The battery can switch from standby mode to charging to discharging in milliseconds. It can also ramp up from

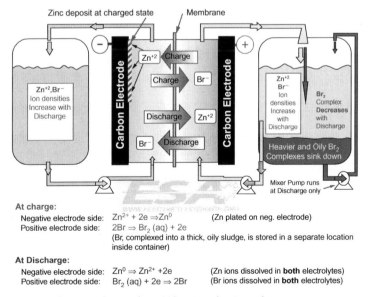

At charge:
Negative electrode side: $Zn^{2+} + 2e \Rightarrow Zn^0$ (Zn plated on neg. electrode)
Positive electrode side: $2Br \Rightarrow Br_2 (aq) + 2e$
(Br, complexed into a thick, oily sludge, is stored in a separate location inside container)

At Discharge:
Negative electrode side: $Zn^0 \Rightarrow Zn^{2+} + 2e$ (Zn ions dissolved in **both** electrolytes)
Positive electrode side: $Br_2 (aq) + 2e \Rightarrow 2Br$ (Br ions dissolved in **both** electrolytes)

FIGURE 4.17 Charging and discharging of zinc–bromine batteries.

					Table 4.1 Batteries for Renewable Energy Systems
Type	**Nominal voltage, V**	**Energy density, wh/L**	**Cycle life**	**Consumer price, $/kw**
Lead acid	2.0	70	250-500	5-8
Nickel cadmium	1.2	60-100	300-700	N/A
Sodium sulfur	2.0	367	2000	N/A
Zinc bromine	1.8	N/A	N/A	N/A
Vanadium redox	1.4–1.6	N/A	>10,000	N/A
Polysulfide bromide	1.7	N/A	N/A	400

full shutdown to full operation in a matter of minutes. Other applications include UPS and power quality (mitigating surges, intermittency).

The power-quality aspect of the zinc–bromine battery makes it a good match for renewable energy systems because they will have some fluctuations of energy output. The zinc gains two electrons that give it a negative charge. When it discharges, the process reverses itself. The chemical equation for the negative electrode reaction when charging, which causes the plating of zinc on the electrode, is as follows:

$$Zn_{(S)} \leftrightarrow Zn^{2+}_{(aq)} + 2e^-$$

The following is the chemical equation for the positive electrode while charging:

$$Br_{2(aq)} + 2e^- \leftrightarrow 2Br^-_{(aq)}$$

The zinc–bromine battery has many uses because of its many different configurations and the fact that its performance is doesn't decline after many cycles. As the ZBB becomes more developed and cost goes down to levels below those of lead-acid batteries, its use is most likely to become more prominent, and its use in load leveling will increase. As a society, we are trying to stop massive use of oil; thus, alternative energy applications will become more prominent, and because of this, use of energy storage such as the zinc–bromine battery is expected to grow.

There are other types of batteries, but they are expensive to use for energy storage. They are used in electric vehicles and other applications. Examples of some are: nickel–metal hydride, lithium-ion, zinc–air, nickel–zinc, nickel–hydrogen, and lithium polymer batteries. More information about the batteries discussed can be found in Table 4.1.

4.2 FUEL CELLS

The fuel cell can trace its roots back to the 1800s. A Welsh-born, Oxford-educated barrister, who practiced patent law and also studied chemistry, or "natural science" as it was known at the time, named Sir William Robert Grove (Figure 4.18) realized

FIGURE 4.18 Photograph of William Grove.

that if electrolysis, using electricity, could split water into hydrogen and oxygen then the opposite would also be true. Combining hydrogen and oxygen, with the correct method, would produce electricity thought Sir Grove. To test his reasoning, he built a device that would combine hydrogen and oxygen to produce electricity—the world's first gas battery. His invention was a success, and Grove's work advanced the understanding of the idea of conservation of energy and reversibility.

4.2.1 Introduction

Fuel cells are a relatively new technology that will reform the way we produce electricity across the world. This type of energy generation is more efficient, lower cost, and a cleaner alternative to today's conventional methods (i.e., coal, gasoline, nuclear, diesel). The name fuel cell was coined in 1889 by Charles Langer and Ludwig Mond, who demonstrated a fuel cell that could develop 6 Amps at 0.73 V. The cell used thin perforated-platinum electrodes. Mond and Langer also worked on fuel cells that converted coal gas directly into electricity. It was at this time, however, that the internal combustion engine began to exert its dominance and fuel cell development took a backseat.

Despite the internal combustion engine, some research on fuel cells did continue. In 1893, Friedrich Wilhelm Ostwald explained the fundamental interactions of the fuel cell. He described how Grove's "gas battery" really worked. Ostwald identified each part of the fuel cell and its function in the reaction.

Modern use of the fuel cell is attributed to the work of Francis Thomas Bacon. Starting his research in the early 1930s, Bacon was able to construct an operational fuel cell with nickel electrodes and alkaline electrolyte in 1932. In 1939, Bacon demonstrated a nickel-gauze electrode, high-pressure fuel cell. However, due to many technical hurdles, it was not until 1959 that Bacon demonstrated a larger system, a 5-kW fuel cell system. Bacon's research was the genesis of the fuel cells used in the NASA space programs; Pratt & Whitney licensed Bacon's work for use in building the fuel cells for the *Apollo* missions.

NASA and the space program provided fuel cells with the initial research and development technology required. Since their adoption by the space program, fuel cell technology has achieved widespread recognition by industry and government as a clean energy source for the future. With this in mind, the amount of interest in fuel cells has expanded exponentially to where 8 of the 10 largest companies in the world are evaluating fuel cells in some respect. Today, billions of dollars have been spent on research and the commercialization of fuel cell products. Over the next couple of years, they will begin to become available to consumers.

4.2.2 Fuel cell components and operation

Fuel cells convert a fuel into electricity using a chemical reaction. By not using any combustion, as internal combustion engines or turbines do, fuel cells are not constrained by conventional thermodynamic efficiency limitations nor do they produce the pollution inherent with the compression and combustion of the fuel and air.

Fuel cells are constructed of several basic parts. The electrolyte provides the medium for the migrating ions and the electrodes, both an anode and cathode, provide an electrical path for the displaced electrons. On or near the electrodes, a catalyst speeds up the reaction by lowering the chemical activation energy. A fuel, usually hydrogen or a hydrogen-rich fuel, is fed to the anode. An oxidizer, almost always oxygen, is fed to the cell at the cathode.

Each type of fuel cell has its own set of reactions. In alkali fuel cells, the migrating charges are hydroxyl ions. Molten carbonate uses carbon trioxide ions. Solid oxide fuel cells have oxygen ions migrating through the solid ceramic electrolyte. In both phosphoric acid and proton-exchange membrane fuel cells, the migrating ions are protons. During the reaction, electrons are released at the anode and collected at the cathode, driving the desired electrical current. Fuel cells also produce heat. Heat can be recovered and made useful using conventional cogeneration techniques. The heat boils water or another working fluid that turns a turbine to generate additional electricity.

There are many types of fuel cells on the market today that have a wide range of operating temperatures, pressures, and different topologies; however, the general mode of operation is similar in each. Figure 4.19 is a graphical representation of a polymer electrolyte membrane (PEM) fuel cell. Hydrogen, the fuel, is injected into the anode of the device and oxygen from the air is injected into the cathode.

The hydrogen reacts with a catalyst, the anode electrode, and has its valance electrons stripped off. These electrons are not able to pass through the electrolyte and are carried through external circuitry to a load (e.g., lamp, motor, vehicle) and ultimately

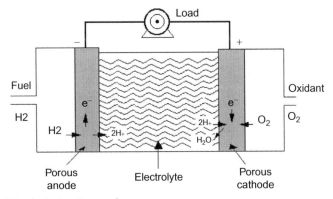

FIGURE 4.19 Basic fuel cell operation.

end up at the unit's cathode electrode. Once the electrons reach the cathode, they again interact with a catalyst and are combined with oxygen from the air and the positively charged hydrogen ions, which passed through the electrolyte to produce H_2O (water). The following will give a clearer representation of what is happening:

At the anode the hydrogen is split

$$H_2 \Leftrightarrow 2H^+ + 2e^-$$

At the cathode the electrons are recombined and water is formed

$$2H^+ + \frac{1}{2}O_2 + 2e^- = H_2O$$

4.2.3 Types of fuel cells

There are at least six common types of fuel cells; they are described in the subsections that follow. Each has its own advantages and disadvantages. More types are currently being developed.

Alkaline fuel cells

Alkaline fuel cells (AFC) are the type developed by Bacon and refined for and used in the space program. This type of requires pure gas inputs, both hydrogen and oxygen. In space, these pure elements are already available and used for propulsion, so the adaptation of there use in the fuel cell was natural. Any impurities in the fuel, such as carbon dioxide or monoxide, will react to form a solid carbonate. This solid carbonate interferes with the chemical reactions in the cell and reduces efficiency and power production. An additional benefit of the AFC is the pure water produced, which on the manned spacecraft was put to good use as drinking water.

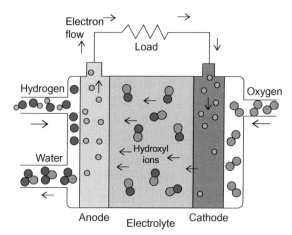

FIGURE 4.20 Operation of the alkaline fuel cell.

The electrolyte of the alkaline fuel cell is potassium hydroxide (KOH) dissolved in water. The electrodes are platinum, an expensive material that also acts as the catalyst for the reaction. The porous catalysts were developed by Bacon in the late 1930s along with the use of pressurized gasses to keep the electrolyte from flooding the electrodes. Figure 4.20 shows the operation of an AFC. Note the pure water being produced at the anode.

The power output range is typically from a few hundred watts to a few kilowatts. In 1959, one of the first practical demonstrations of fuel cells used alkaline technology. The farm implement maker Allis-Chalmers made a farm tractor powered by a 1008 cell AFC stack, which powered it with about 15 kW of power, enough to pull 3000 pounds.

Although alkaline fuel cells are the most temper mental of all fuel cells, it can produce the maximum amount of energy (80% efficiency) when used as a water-heating device. AFCs use an KOH electrolyte because it is the most conductive of all alkaline hydroxides; however, this requires extremely pure hydrogen and oxygen input to avoid poisoning. The cell cannot internally reform any fuel because of the 80 °C cell-operating temperature, so this requires additional equipment to operate unless pure hydrogen is stored in a tank.

Hydrogen, at the anode, reacts with the electrolyte, creating water and two electrons that both meet at the cathode with oxygen to complete the circuit where it is combined (see Figure 4.20). The electrolyte constantly flows through the cell, providing cooling by convection of the porous (and catalyzed) graphite electrodes from which it picks up hydroxyl ions and a small amount of water in the process. In an actual cell, one-third of the water produced drains on the cathode side while two-thirds resides on the anode side.

Again, because of the liquid nature of the electrolyte, semipermeable, Teflon-coated carbon material is used as electrodes, which are heavily catalyzed as compared with other types of fuel cells because of the low operating temperature. Typical cell voltages at $70\,^\circ$C are 0.878 V with a current of 100 mA/cm^2. The chemical reaction that occurs is as follows:

At the anode

$$H_2 + 2OH^{..} \rightarrow 2H_2O + 2e^{..}$$

At the cathode

$$O_2 + 2H_2O + 4e^{..} \rightarrow 4OH^{..}$$

After recombination

$$2H_2 + O_2 \rightarrow 2H_2O + \text{electric energy} + \text{heat}$$

Molten-carbonate fuel cells

Molten-carbonate fuel cells (MCFC) operate at a much higher temperature than the earlier AFCs. The MCFCs use inexpensive nickel catalysts as the electrodes. Initial MCFC development really started in the 1950s with Dutch scientists Broers and Ketelaar. They were the first to use molten carbonate as the electrolyte. By the 1960s, the US Army, with Texas Instruments, had developed MCFC units that could operate, with a reformer, from gasoline.

The MCFC is uniquely suited to operation from carbon-based hydrogen-rich fuels. The reaction using the CO_3 produces carbon dioxide. Thus, along with the oxygen provided to the cathode, carbon dioxide must also be provided. See Figure 4.21 for an operational schematic of the MCFC.

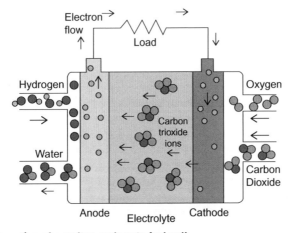

FIGURE 4.21 Operation of a molten-carbonate fuel cell.

MCFC stacks operate at a high temperature of about 650 °C. The high temperature makes cogeneration of electricity from waste heat obvious, thus leading to higher efficiencies. MCFC technology is usually best suited for nonmobile, relatively constant use applications—usually utility-scale electric generation. The molten-carbonate fuel cell, which operates at 600 °C, can use CO as the fuel input on the cathode side but needs hydrogen on the anode side.

Carbonate ions are produced at the cathode and flow across the membrane to react with hydrogen and form two electrons. The temperature is high enough for additional power production through cogeneration of steam and low enough to eliminate the need of expensive catalysts required in the solid oxide fuel cells (SOFC). A MCFC operates nominally at 0.16 A/cm^2 and 0.75 V per cell with increased performance when pressurized conditions exist.

Nickel compounds are used for the electrodes while the electrolyte contains a mixture of 68% lithium carbonate and 32% potassium carbonate in a porous gamma–lithium–aluminum–oxide matrix. The efficiency using this system has risen to 50% in a combined cycle, which is one where steam is produced from the operating temperature of the device; the steam is also used in a productive manner earlier referred to as cogeneration. In addition, MCFC, like the SOFC, is used for mW-size power plants because of its heat generation.

At the anode

$$H_2 + CO_3^= \rightarrow H_2O + CO_2 + 2e^-$$

At the cathode

$$\tfrac{1}{2}O_2 + CO_2 + 2e^- \rightarrow CO_3^=$$

After recombination

$$H_2 + \tfrac{1}{2}O_2 + CO_2(\text{cathode}) \rightarrow H_2O + CO_2(\text{anode})$$

Phosphoric acid fuel cells

Phosphoric acid fuel cells (PAFC) are similar to the sulfuric acid-based cells developed by William Grove. In 1962 in their paper, Elmore and Tanner proposed the use of "intermediate temperature fuel cells" that used phosphoric acid as the electrolyte. In the 1960s, the US Army explored the use of a PAFC fueled by common fuels, namely diesel and gasoline. This fuel cell can tolerate limited concentrations of carbon monoxide but cannot tolerate sulfur, as it will damage the electrode. See Figure 4.22 for an operational schematic of the PAFC.

This fuel cell runs at a moderate temperature of between 150 and 200 °C. Due to the high concentrations of the phosphoric acid, all internal components must be very corrosion-resistant and the electrodes are platinum. Phosphoric acid fuel cells are

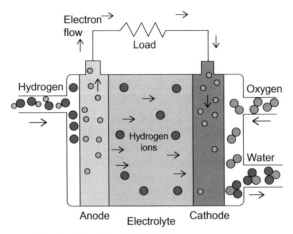

Electron flow

Load

Hydrogen

Oxygen

Hydrogen ions

Water

Anode   Electrolyte   Cathode

FIGURE 4.22 An operational schematic of the PAFC.

employed in transit buses and building-scale electrical generation. PAFCs are the oldest type with origins that extend back to the creation of the fuel cell concept.

Many different acids, such as sulfuric, have been used to boost performance but when the temperature increases above 150 °C, high rates of oxygen reduction, which enables phosphoric acid to perform best, are possible. The temperature allows the cell to tolerate 1 to 2% CO (cobalt) and a few parts per million (ppm) of sulfur in the reactant stream. This benefits the steam-reforming process by reducing the requirement of pure hydrogen input to the anode. The heat generated is not enough for cogeneration of steam but is able to warm water and to act as a heater for an increased overall efficiency. The electrolyte is surrounded by porous graphite carbon coated with Teflon to allow gases to the reaction sites but not allow the liquid electrolyte out.

The efficiency of this system is much lower than that of other systems at 40%, but because of its history, it can be controlled well. PAFCs are the only fuel cells in commercial production and have more than 75 MW of demonstrated use around the world with the largest single plant being at 11 MW. The chemical reaction that takes place is as follows:

At the anode

$$H_2 \rightarrow 2H^+ + 2e^-$$

At the cathode

$$\frac{1}{2}O_2 + H_2 \rightarrow H_2O$$

After recombination

$$\frac{1}{2}O_2 + 2H^+ + 2e^-$$

Solid oxide fuel cells

With a very high-operating temperature, solid oxide fuel cells (SOFC) have the distinct advantage of being able to operate with hydrogen-rich fuels without the use of a reformer. The high temperature makes cogeneration viable, if not necessary, for best efficiency. Solid oxide fuel cells operate with a ceramic, most commonly a mixture of zirconium oxide and calcium oxide. Figure 4.23 shows the operational schematic of a SOFC.

Historically, research into solid oxide fuel cells has been stunted by the difficulty of finding a temperature-tolerant, electrical-conductive ceramic that is porous enough to permit the oxygen ions to migrate between the electrodes. SOFCs operate at 1000 °C with efficiencies up to about 60%.

Solid oxide fuel cells have become of more interest lately due to their high-efficiency and compact, low-maintenance design. Currently, more than 40 companies are pursing research on the SOFC to supplement or replace current electricity-generation plants. The scope of them is scaleable to meet the demands of a small building to that of utility-scale power. The US Department of Energy (DOE) and Siemens Westinghouse are planning a 1-MW station.

SOFCs, which operate at the highest temperature of all fuel cells ($> 800 °C$) are not the most reactive because of the low conductivity of its ionic-conduction electrolyte (yttria-stabilized zirconium). Carbon monoxide (CO) and hydrocarbons, such as methane (CH_4), are the preferred fuels to use in SOFCs. The water–gas shift involving $CO(CO+H_2O \rightarrow H_2+CO_2)$ and the steam reforming of $CH_4(CH_4+H_2O \rightarrow 3H_2+CO)$ occur at the high-temperature environment of the SOFC to produce H_2, which easily oxidized at the anode and eliminates the need for an internal reformer (Figure 4.24).

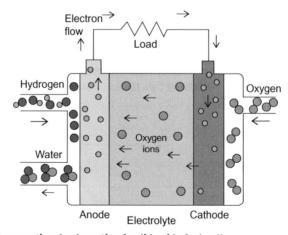

FIGURE 4.23 An operational schematic of solid oxide fuel cell.

FIGURE 4.24 A 220 kW Westinghouse SOFC with a microturbine.

The electrolytes used in all fuel cells, except molten carbonate, have to have certain properties that allow transmission of the hydrogen gas and present resistance to electrical current, which forces the electrons that are stripped off of the hydrogen at the anode to follow the path of least resistance and to flow to the external circuit. Many advances have been made in solid oxide fuel cell research to increase the chemical-to-electrical efficiency to 50%; however, because of the conductivity and the heat required, it has been used mainly in large power plants that use the cogeneration of steam for additional power. The steam is produced when the hydrogen, oxygen, and electrons are recombined at the cathode.

Because of the high temperature, the cell requires no expensive catalysts, or additional humidification and fuel treatment equipment, which excludes the cost of these items. The primary drawback to this type of fuel cell is the cost of the containment due to the intense temperatures. The chemical operation that occurs is as follows:

At the anode

$$H_2 + O^= \rightarrow H_2O + 2e^-$$

At the cathode

$$\tfrac{1}{2}O_2 + 2e \rightarrow O$$

After recombination

$$H_2 + \tfrac{1}{2}O_2 \rightarrow H_2O$$

Proton exchange membrane fuel cells

A most promising technology, proton exchange membrane (PEM) fuel cells operate with a thin permeable polymer for an electrolyte. This material, a Teflon derivative, is solid, thus simplifying the construction of the cell. See Figure 4.25 for the schematic of the PEM cell. The PEM cell has a spotted history.

Originating from work at General Electric during the 1960s for the US Navy, the cell was designed to be fueled by the hydrogen generated from the reaction of water and lithium hydride. An easily portable formula for the users, the solution was hindered by the high cost of the platinum electrodes. NASA investigated the use of the PEM technology to power the Gemini project but did not get it installed until *Gemini 5*. Prior flights relied on batteries instead. Proton-exchange cells were left behind with the Apollo project as the engineers chose the alkaline fuel cells.

Research by Texas A&M University and the Los Alamos National Labs in the late 1980s and early 1990s produced a reduction in the amount of platinum required for operation, thus opening the door to more economical use of the fuel cell. PEM cells and their derivatives are considered very promising for use in small-to-medium–scale power production. Automakers are investigating using proton exchange fuel cells to provide electricity to power future electric vehicles. (See also Figure 4.26.)

PEM cells operate at around 80 °C, like AFCs, but do not have strict requirements on the fuel input, or the efficiency of the alkaline, but at 60% efficiency, it comes in second only to alkaline. The per fluorinated sulfuric acid membrane is sandwiched between two platinum-catalyzed porous electrodes and compressed using bolts. The electrolyte is very sensitive to CO contamination, so the inlet streams must be purified. Air can be used instead of oxygen as a reducing agent, but the efficiency of the system falls drastically as the CO content in the cathode rises above 0.17%.

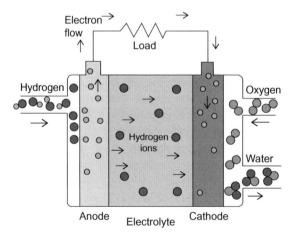

FIGURE 4.25 A schematic of the PEM cell.

FIGURE 4.26 Ballard, a leader in PEM technology, introduced the Nexa™ Power Module in 2001.

The temperature and the solid type of membrane make mobile power generation possible, which makes the use of this type of fuel cell available to alternative energy vehicles. The chemical reaction taking place within the proton exchange membrane fuel cell is as follows:

Anode: $H_2 \rightarrow 2H + 2e^-$
Cathode: $\frac{1}{2}O_2(g) + 2H + 2e^- \rightarrow H_2O$
Cell: $H_2 + \frac{1}{2}O_2 \rightarrow H_2O$

Protonic ceramic fuel cells

This new type of fuel cell is based on a ceramic electrolyte material that exhibits high-protonic conductivity at elevated temperatures. Protonic ceramic fuel cells (PCFC) share the thermal and kinetic advantages of high-temperature operation at 700 °C with molten carbonate and solid oxide fuel cells, while exhibiting all the intrinsic benefits of proton conduction in polymer electrolyte and phosphoric acid fuel cells. The high operating temperature is necessary to achieve very high electrical fuel efficiency with hydrocarbon fuels.

PCFCs can operate at high temperatures and electrochemically oxidize fossil fuels *directly* to the anode. This eliminates the intermediate step of producing hydrogen through the costly reforming process. Gaseous molecules of the hydrocarbon fuel are absorbed on the surface of the anode in the presence of water vapor, and hydrogen atoms are efficiently stripped off to be absorbed into the electrolyte, with carbon dioxide as the primary reaction product. Additionally, PCFCs have a solid electrolyte, so the membrane cannot dry out as with PEM fuel cells, or liquid cannot leak out as with PAFCs. Protonetics International, Inc., is the primary company researching this type of fuel cell (Figure 4.27).

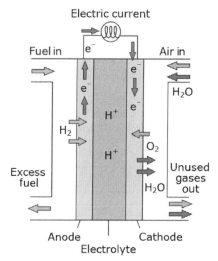

FIGURE 4.27 A PCFC at Protonetics International, Inc.

The efficiency of the PCFC is targeted to be around 60% with a pipeline of direct natural gas. The chemical reaction taking place within the protonic cermaic fuel cell is as follows:

Anode: $CH_4 + 2H_2O \rightarrow CO_2 + 8e^- + 8H^+$
Cathode: $8H^+ + 8e^- + 2O_2 \rightarrow 4H_2O$

Direct methanol fuel cells

These cells are similar to the PEM cells in that they both use a polymer membrane as the electrolyte. However, in the direct methanol fuel cell (DMFC), the anode catalyst itself draws the hydrogen from the liquid methanol, eliminating the need for a fuel reformer (Figure 4.28). Efficiencies of ~40% are expected with this type of cell,

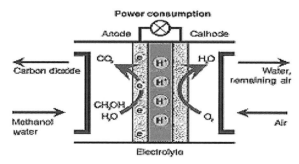

FIGURE 4.28 Example of direct methanol fuel cells.

Source: Courtesy of Smart Fuel Cells

which would typically operate at a temperature between 50 and 100 °C. Better efficiencies are achieved at higher temperatures. The chemical reaction taking place within the direct methanol fuel cell and some applications are as follows:

On the anode: $CH_3OH + H_2 = CO_4 + 6H^+ + 6e$
On the cathode $6H^+ + 6e + 3/2\ O_2 = 3H_2O$
Applications: EV, military, cellular phones, laptop computers

Regenerative fuel cell

A refinement of the PEM fuel cell technology holds very interesting promise for future use. If constructed properly, a variant of the PEM can be made to function in reverse. This cell, called a regenerative fuel cell (RFC), takes water and electricity and makes hydrogen. The external power source is not available, so the hydrogen can be consumed and water and electricity produced. See Figure 4.29 for more details on the operation of the RFC.

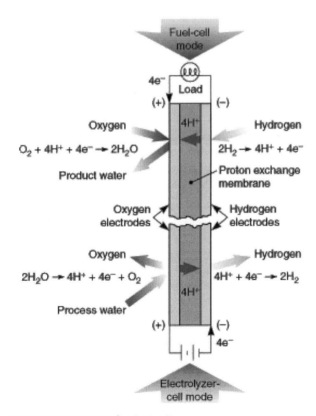

FIGURE 4.29 Example of a regenerative fuel cell.

With a RFC, an automobile can be made to be self-fueling. Simply plug in your car at home at night when the electricity rates are low and "fill up the tank." This innovation frees the automobile from needing a hydrogen distribution system because it can use the electrical distribution system that already exists. The regenerative fuel cell can be used in conjunction with a house's solar or wind power. When the excess electricity is available, you can make hydrogen. The Sun is down and the wind is quiet the RFC reverses and produces the needed electricity.

4.2.4 Comparison of the main types of fuel cells

Molten carbonate (MCFC) and solid oxide (SOFC) fuel cells require high-operating temperatures at all times, so almost certainly rule out these fuel cell types for propulsion of cars. Alkaline fuel cells (AFC) are capable of good power density, but they cannot tolerate even very low concentrations of carbon dioxide, which is a major constituent of the processed fuel. In automotive applications, it would be impractical to remove carbon dioxide completely from the processed fuel.

The phosphoric acid fuel cell (PAFC), currently being commercialized for stationary power applications, can use ambient air and processed fuel but has only modest power density, a deficiency that translates into relatively large volume and weight, as well as higher cost. Another disadvantage is that the PCFC cannot generate power at ambient temperature but must be preheated to at least $100\,^{\circ}C$ before current can be drawn. Meeting the automotive requirement for rapid start-up would, therefore, be very difficult.

The direct methanol fuel cell (DMFC) is the only practical carbonaceous fuel that has significant electrochemical reactivity at the fuel cell electrodes in the temperature range of interest for automobile applications. However, the DMFC has been hampered by two major problems: poor performance (current density below the levels required for automobiles) and rapid diffusion of methanol through the membrane to the air electrodes where it is readily oxidized in an electrochemical "short-circuit" reaction. This methanol "crossover" problem not only reduces fuel-utilization efficiency substantially (typically by 30% or more) but also depresses the potential of the air cathode, thus the cell voltage, thereby causing an additional loss of energy efficiency.

When considering applicability of fuel cells for automobile propulsion, the important factors are operating temperature and power density; being compatible with processed fuel and air; starting up quickly and responding rapidly to frequent load changes; resisting shock and vibration; and being relatively easy to control and maintain (Table 4.2). Considering all these factors, the proton exchange membrane (PEM) fuel cell is most likely to meet the requirements.

The PEM fuel cell can operate at a relatively low temperature of $\sim\!80\,^{\circ}C$, can use air, and has excellent performance with hydrogen is used as the fuel. This fuel cell uses a thin plastic sheet as its electrolyte, which is easy and safe to handle in manufacturing and in later use. Unlike some other electrolytes, their solid plastic membrane can tolerate a modest pressure differential across the cell, making for easy

Table 4.2 Characteristics of Various Types of Fuel Cells

| Type | Type of electrolyte | Operating temperature (°C) | Current density | Need for fuel processor | Compatibility with CO_2 | Stage of development | Current prospects for | |
							High efficiency	Low cost
PEMFC	Proton exchange membrane	70-80	High	Yes	Yes	Early prototypes	Good	Good
AFC	Alkaline	80-100	High	Yes	No	Space application	Good	Good
PAFC	Phosphoric acid	200-220	Moderate	Yes	Yes	Early commercial applications	Good	Fair
MCFC	Molten carbonate	600-650	Moderate	Yes[a]	Yes	Field demonstrations	Good	Fair
SOFC	Solid-oxide	800-1000	High	Yes[a]	Yes	Laboratory demonstrations	Good	Fair to good
DMFC	Proton exchange membrane	70-80	Moderate	No	Yes	Research	Poor	Poor to fair
PCFC	Ceramic	700	High	No	Yes	Research	High	High

pressurization, which increases power density, simplifies the rest of the system, and reduces cost. Not surprisingly, all ongoing programs developing automotive fuel cell technology and systems are concentrating their efforts on the PEM technology. Thus, we will also concentrate our discussion about fuel cell technology on the PEMs for electric vehicle applications.

4.2.5 Fuel cell efficiency

When considering a fuel cell for an application, we must consider how efficient the fuel cell will be (Table 4.3). Fuel cell efficiency directly impacts the operating cost as well as fuel storage requirements. In mobile applications, the efficiency becomes even more important because the fuel must be moved with the cell. A less efficient fuel cell must move fuel to travel the same distance, thus wastes some of the generated energy moving the additional fuel.

It is important to notice that the efficiency range for the fuel cells rang from 40 to 80%. One major advantage of fuel cells becomes apparent when we compare their efficiencies to the efficiency of the internal combustion engine. The typical efficiency of an internal combustion engine is on the order of 20%. Thus, to power a vehicle with the same performance and range, we only need to carry half as much fuel, assuming a 40% efficient PEM cell stack.

4.2.6 Applications of fuel cells

Fuel cells are being employed in numerous applications in everything from cellular phones to prime power generation for utility companies. The application of them can be divided into six separate categories: stationary, residential, portable power, landfill/wastewater treatment, transportation, and power plants generating electricity.

Stationary

Hundreds of fuel cell systems have been installed all over the world in hospitals, nursing homes, hotels, office buildings, schools, and at airport terminals; they provide primary power or backup. In large-scale building systems, fuel cells can reduce facility energy service costs by 20 to 40% over conventional energy service. United Technologies Company (UTC) has developed a 200 kW stationary fuel cell. the PC25™ (Figure 4.30). Four of them have been installed in the First National Bank of Omaha

Table 4.3 Fuel Cell Efficiency Values	
Fuel Cell Type	**Efficiency**
Alkaline	70%
Molten carbonate	60–80% with cogeneration
Phosphoric acid	40–80% with cogeneration
Solid oxide	60%
Proton exchange membrane	40–50%

Photo Courtesy of UTC

FIGURE 4.30 The 200 kW stationary fuel cell, the PC25.

as a backup supply in the event of a power outage. Heat from the chemical reaction is reused for space heating, thus bringing the overall efficiency to more than 80%.

Residential

Fuel cells are ideal for power generation, either connected to the electric grid to provide supplemental power and backup assurance for critical areas, or installed as a grid-independent generator for onsite service in areas that are inaccessible by power lines. Since fuel cells operate silently, they reduce noise pollution, as well as air pollution, and the wasted heat from a fuel cell can be used to provide hot water or space heating.

There are three main components in a residential fuel cell system: the hydrogen fuel reformer, the fuel cell stack, and the power conditioner. Many of the prototypes being tested and demonstrated extract hydrogen from propane or natural gas. (See Figure 4.31.) The fuel cell stack converts the hydrogen and oxygen from the air into electricity, water vapor, and heat. The power conditioner then converts the electric DC current from the stack into AC current on which many household appliances operate. Asia Pacific Fuel Cell Technologies, Ltd. (APFCT) estimates the expected payback period on a residential fuel cell for a typical homeowner to be four years. The initial price per unit in low-volume production will be approximately $1500/kW. Once high-volume production begins, the price is expected to drop to $1000/kW, with the ultimate goal of getting costs below $500/kW. Fuel cell developers are racing to reach these cost targets.

Many other companies are developing and testing fuel cells for residential applications, working together with utilities and distributors to bring them to market. Even automakers, such as GM and Toyota, are branching beyond vehicles and spending money on research and development for stationary applications.

To promote the commercialization of residential fuel cells, a bill proposing a stationary fuel cell tax credit was introduced early this decade in both the House

FIGURE 4.31 A 5 kW SOFC residential fuel cell.

Source: Courtesy of Plug Power, Inc.

(H.R. 1275) and the Senate (S. 828) and referred to the Ways and Means Committee and the Finance Committees, respectively. If approved it would have allowed US businesses and residential taxpayers a $1000 per kW credit toward the purchase of fuel cell systems for stationary commercial and residential applications. It was to be available for five years, starting on January 1, 2002, through December 31, 2006, after which fuel cell manufacturers were expected to produce a product at market entry cost. On top of that, many states have net metering laws in place that allow qualified customers to sell surplus electricity and heat produced from a fuel cell qualifies back to the grid.

Portable power

Miniature fuel cells, once they become available to the commercial market, will let consumers talk for up to a month on a cellular phone without recharging. Fuel cells will change the telecommuting world, powering laptops and cellular devices hours longer than batteries. Other applications for micro-fuel cells include doctors' pagers, video recorders, portable power tools, and low-power remote devices such as hearing aids, smoke detectors, burglar alarms, hotel locks, and meter readers. The miniature fuel cells generally run on methanol, an inexpensive wood alcohol also used in wiper fluid. Examples of portable devices powered by fuel cells are shown in Figures 4.32. through 4.35.

Landfill/wastewater treatment

Fuel cells currently operate at landfills and wastewater treatment plants across the country, proving themselves as a valid technology for reducing emissions and generating power from the methane gas they produce. In Oregon, the city of Portland was looking for a clean, efficient way to power its water treatment facility

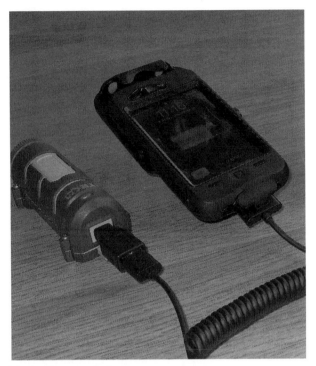

FIGURE 4.32 A fuel cell-powered phone.

Source: Courtesy of Raul Gonzalo, Creative Commons License.

FIGURE 4.33 A fuel cell-powered flashlight.

(Figure 4.36). The city selected UTC to install one of its 200-kW fuel cells (PC25) that convert anaerobic digester gas generated by the facility into usable heat and electricity for it. By using waste gas that might otherwise be flared, the project makes use of a free source of fuel. Additionally, the electricity generated is considered a renewable source of energy.

FIGURE 4.34 A fuel cell-powered laptop computer.

FIGURE 4.35 A fuel cell powered television.

Photo Courtesy of UTC

FIGURE 4.36 Harnessing the wastewater treatment gas with a fuel cell.

Transportation

Fuel cells are poised to overtake the Internal Combustion Engine (ICE). All the major automotive manufacturers have a fuel cell vehicle either in development or in testing right now: Honda, Toyota, DaimlerChrysler, GM, Ford, Hyundai, and Volkswagen. Figure 4.37 shows a fuel cell electric car.

Fuel cells are a promising technology that may gradually replace combustion engines and chemical batteries as the primary energy sources for much of society. Fuel cells are used to generate electricity on a utility scale. They power transit busses, as shown on left in Figure 4.38. Fuel cells show promise to be widely used to power personal vehicles, be they cars, scooters, or airplanes; Figure 4.39 shows a fuel cell-powered airplane. Fuel cells may also show up in more unexpected places.

Power plants generating electricity for utility companies

A fuel cell power plant consists of the following three parts:

PEM fuel cell stacks: Most fuel cells produce less than 1.16 V of electricity—far from enough to generate electricity as a power plant. Therefore, multiple cells must be assembled into a fuel cell *stack*. To deliver the desired amount of energy, the cells can be combined in series and parallel circuits; series assembly yields higher voltage and parallel allows a stronger current to be drawn. The potential

FIGURE 4.37 A fuel cell-powered electric vehicle.

FIGURE 4.38 A fuel cell-powered electric bus.

FIGURE 4.39 PEM fuel cell-powered airplane.

Source: Courtesy of AeroVironment, Inc.

power generated by a fuel cell stack depends on the number and size of the individual cells that comprise the stack and the surface area of the PEM.

Fuel processor: Some kind of a reformer is used to produce hydrogen-rich gas for use within the fuel cell stack. There are many kindsof reformers that can be used to provide the hydrogen.

Balance of plant: Beyond the two major functional subsystems (stack and fuel processor) a fuel cell power plant has numerous supporting and auxillary components called the balance of plant. This includes, the computer-based control system, the power conditioner, the air-handling system (i.e., compressor–expander–drive motor assembly), and the humidifier system.

Figure 4.40 shows a schematic diagram of a fuel cell power plant.

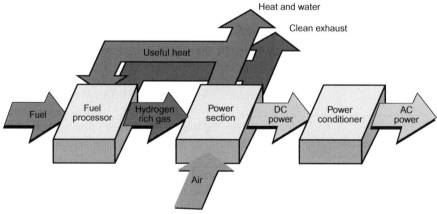

FIGURE 4.40 Schematic representation of a fuel cell power plant.

4.2.7 **Future outlook for fuel cells**

Over the coming decades, concerns about depleting stocks of natural resources and a growing awareness of the environmental damage caused by widespread burning of fossil fuels will help to drive the development of fuel cells for both transport and stationary power sources. In the past, they were always large and extremely expensive to manufacture. However, with the interest and financing of government labs (e.g., Los Alamos National Lab) and private entities, the fuel cell is quickly becoming economically viable and physically practical.

The primary reason that fuel cells have not become more widespread is cost. Fuel cells developed for the space program during the 1960s and 1970s were extremely expensive and impractical for terrestrial power applications. The past three decades have seen significant efforts made to develop more practical and affordable designs for stationary power applications. But, progress has been slow. Today, the most widely marketed fuel cells cost about $4500/kW; by contrast, a diesel generator costs $800 to $1500/kW to operate, and a natural gas turbine can cost even less. Recent technological advances, however, have significantly improved the economic outlook for fuel cells.

The development of improved, less-expensive conducting materials now allow the newest fuel cells to work at higher temperatures and generate more power per volume (i.e., more "power density"), both of which lower costs. Higher operating temperatures also give newly developed fuel cells the capability to produce exhaust heat that can be used for space and water heating or even additional power production. Techniques also have been developed to separate hydrogen from natural gas inside the fuel cell (i.e., "internal reforming"), eliminating the expense of a separate system.

Costs, however, are still not competitive with more conventional power sources. State-of-the-art fuel cells now being tested are likely to cost around $1200/kW— comparable to a large-scale coal-fired power plant but still too expensive for most onsite (i.e., "distributed") power uses. The DOE has launched a major initiative to bring about dramatic reductions in fuel cell costs.

In more recent decades, a number of manufacturers, including major automakers and various federal agencies, have supported ongoing research into the development of fuel cell technology for use in fuel cell vehicles (FCV) and other applications. Fuel cell energy is expected to replace traditional power sources in coming years—even micro fuel cells will be used in cell phones.

4.2.8 **Reformers**

As was mentioned earlier, the quality of the fuel and oxygen required for operation of fuel cells varies from unit to unit. For this reason, there is a need for an element of the overall system called a *reformer*. Reformers are used to extract hydrogen from a hydrogen-rich fuel. There are two principal methods of reforming hydrogen-rich, carbon-based fuels: steam and partial oxidation. Steam-reforming systems tend to be large and require a lengthy startup time but have good efficiencies. These attributes make them attractive for use in stationary fuel cell applications.

With a quick startup time and decent efficiency a noncatalytic, partial oxidation reformer may be suitable for providing hydrogen to use in mobile fuel cell systems. These systems do not necessary produce pure hydrogen of the type demanded by the alkaline-type fuel cells, but they may be engineered to produce sufficiently low enough levels of contaminants, mainly carbon monoxide, to power the other types of fuel cell systems.

Applications for reformers

The applications that are being implemented and researched are widespread. There are stand-alone systems where the only source of electricity is the fuel cell in remote areas such as a repeater tower for cell phones or in the case of the space shuttle. There are systems that will be used as portable generators, such as the ones that were designed by Sanyo electronics, which have a capacity to deliver 1 kW for 3 hours and only run at 40 dB of noise.

Electric vehicles will probably be the most lucrative market to apply this new technology in the near future. Reduction of smog and fossil fuel use has become a requirement and will drive the funding of government programs. The US government has already taken steps to require that a certain percentage of vehicles on the road must be electric—why not have the source of that power be fuel cells?

The US Army's small fuel cell development program is just one of many federally funded programs that actively investigate the feasibility of using fuel cells and the acceptable applications for this new technology. One of the major roadblocks for the fuel cell powered vehicle is the development of a safe method of refueling the systems with the hydrogen; they need to operate but since the *Hindenburg* disaster people have been weary of handling hydrogen.

Reformer technology

Reformers were among the many approaches attempted with fuel cells to help solve the problem associated with road traffic in terms of reducing the quantity and improving the quality of emissions. They aid in solving the low-power and high-cost problems associated with fuel cell powered electric vehicle technology. Both methods involve the isolation of hydrogen used as the fuel source for compact hydrogen production system (CHYPS). The three distribution concept solutions to providing hydrogen to fuel cell-powered electric cars are stationary reformers (off-board), mobile (onboard), and hydrogen storage.

Stationary reformers are intended to provide refueling infrastructures, analogous to gas stations. In this scenario, hydrogen is produced at the refueling station, and then transferred to a vehicle with a direct hydrogen system. An infrastructure of this kind could offer the benefits of low-scale hydrogen production and allow for simpler, less costly vehicle systems. Inexpensive, compact, preferably low-pressure, onboard hydrogen storage would still be desirable for this option. Developing such gas-station-like infrastructures would be very expensive and could still pose consumer safety concerns. In essence, the obstacle faced by current fuel cell technology is fuel infrastructure. The alternative would be for vehicles to have hydrogen storage canisters.

The third solution is mobil reformers. A reformer-based vehicle system has the advantage of using the more readily available fuels. There are several ways to produce hydrogen from these more readily available carbonaceous fuels. For example, methanol, natural gas, liquid petroleum gas, gasoline, diesel, coal, biomass, or different alcohols can be used for hydrogen production. The main processes are steam reforming, partial oxidation, gasification, and thermal or catalytic cracking. There is a growing interest in gasoline reformers but this technology is still in its infancy along with the challenge of delivering hydrogen cleanly and efficiently. In essence, the various fuel systems in design must trade-off system complexity and cost versus ease of fueling.

Hydrogen is the lightest element; its properties are shown in Table 4.4. It is by far the most abundant element in the universe and makes up about 90% of it by weight. Hydrogen, as water (H_2O), is absolutely essential to life and is present in all organic compounds. Hydrogen gas was used in lighter-than-air balloons for transport but is far too dangerous because of the fire risk (e.g., *Hindenburg*). The lifting agent for the ill-fated balloon was hydrogen rather than the safer helium. A recent study has shown that the paint used to seal the cloth structure of the *Hindenburg* was the cause of the ignition not the hydrogen as previously thought. Hydrogen can be a safe clean fuel source if handled and/or stored properly.

Hydrogen can be stored in a pure gaseous state at high pressure. It may also be stored in hydrogen-containing chemical compounds. When stored in this manner, it can be extracted via hydrogen reformers. The petroleum and chemical industries have been using reformer technologies for many years to produce not only hydrogen but also a series of other chemicals, including low-emission fossil fuels for use in cars, power generation, and a host of other purposes.

Here is a brief summary of the isolation of hydrogen. In the laboratory, small amounts of hydrogen gas may be made by the reaction between calcium hydride and water.

$$CaH_2 + 2H_2O \rightarrow Ca(OH)_2 + 2H_2$$

Table 4.4 Properties of Hydrogen

Name	Hydrogen
Symbol	H
Atomic number	1
Atomic weight	1.00794 gm
Group number	1
Period number	1
Color	Colorless

This is quite efficient in the sense that 50% of the hydrogen produced comes frrm water. Another very convenient laboratory-scale experiment follows Boyle's early synthesis—the reaction of iron filings with dilute sulphuric acid:

$$Fe + H_2SO_4 \rightarrow FeSO_4 + H_2$$

There are many industrial methods for the production of hydrogen and what is used will depend on local factors such as the quantity required and the raw materials at hand. Two processes in use involve heating coke with steam in the water–gas shift reaction or hydrocarbons such as methane with steam:

$$CH_4 + H_2O(1100°C) \rightarrow CO + 3H_2$$

$$C(coke) + H_2O(1000°C) \rightarrow CO + H_2$$

In both these cases, further hydrogen may be made by passing the CO and steam over hot (400 °C) iron oxide or cobalt oxide:

$$CO + H_2O \rightarrow CO_2 + H_2$$

The transition to clean, highly efficient fuel cell vehicles can be made easier if the early generations of these vehicles can be fueled by gasoline and other fuels that have an existing supply infrastructure. Widespread use of fuel cell vehicles will result in decreased pollutant emissions and increased fuel economies (more miles per gallon). Fuel flexibility will speed the introduction of the technology. The gasoline for fuel cell vehicles may be easier to produce because the fuel cells' performance is not related to octane rating. Today's gasoline is blended with expensive, synthetic octane-enhancing additives such as oxygenates. Reformers will provide an economical hydrogen fuel cell fuel supply. Smaller point-of-use reformers will result in a dedicated production system located at the customer's site. Continued technology development will result in lower fuel costs.

The major reformer technologies

The major reformer technologies available today are steam methane, partial oxidation, auto-thermal, methanol steam, and thermo-catalytic cracking of methane. Several newer experimental technologies are available: sorbent enhanced, ion transport, plasma reformers, and microchannel reformers.

Steam methane reformers. There are a couple of types of steam reformers, one reforms methanol and the other reforms natural gas. When reforming methanol, the molecular formula for methanol is CH_3OH. The goal of the reformer is to remove as much of the hydrogen, H, as possible from this molecule while minimizing the emission of pollutants such as CO. The process starts with the vaporization of liquid methanol and water; heat produced in the reforming process is used to accomplish this. This mixture of methanol and water vapor is passed through a heated chamber that contains a catalyst.

As the methanol molecules hit the catalyst, they split into carbon monoxide and hydrogen gas, H_2:

$$CH_3OH \Rightarrow CO + 2H_2$$

The water vapor splits into hydrogen gas and oxygen. This oxygen combines with the CO to form CO_2. In this way, very little CO is released, as most of it is converted to CO_2:

$$H_2O + CO \Rightarrow CO_2 + H_2$$

In the reforming process for natural gas, which is composed mostly of methane CH_4, the natural gas is processed using a similar reaction. The methane in the natural gas reacts with water vapor to form carbon monoxide and hydrogen gases:

$$CH_4 + H_2O \Rightarrow CO + 3H_2$$

Just as it does when reforming methanol, the water vapor splits into hydrogen gas and oxygen, the oxygen combining with the CO to form CO_2.

$$H_2O + CO \Rightarrow CO_2 + H_2$$

Neither of these reactions is perfect; some methanol or natural gas and carbon monoxide make it through without reacting. These are burned in the presence of a catalyst, with a little air to supply oxygen. This converts most of the remaining CO to CO_2, and the remaining methanol to CO_2 and water. Various other devices may be used to clean up any other pollutants, such as sulfur, that may be in the exhaust stream. It is important to eliminate the CO from the exhaust stream for two reasons:

- If the CO passes through the fuel cell, the performance and life of it are reduced.
- CO is a regulated pollutant, so cars are only allowed to produce small amounts of it.

Partial oxidation reformers. Partial oxidation (POX) reforming of hydrogen is commercially available. It derives the hydrogen from hydrocarbons. Methane or some other hydrocarbon is oxidized to produce carbon monoxide and hydrogen:

$$CH_4 + 1/2O_2 \rightarrow 2\,H_2 + CO$$

The reaction is exothermic and no indirect heat exchange is need. Catalysts are not required because of the high temperature. However, the hydrogen yield can be increased significantly by the use of catalysts. A hydrogen plant based on POX includes a partial oxidation reactor, followed by a shift reactor and hydrogen purification equipment. A partial oxidation reactor is more compact than a steam reformer because it does not need a heat exchanger.

The efficiency of a POX unit is relatively high at 70 to 80%. However, POX systems are typically less energy efficient than steam reforming because of the higher temperatures involved. This exacerbates heat losses and the problem of heat recovery. In the steam plant, heat can be recovered from the flue gas to raise steam for the reaction.

Autothermal reformers. In the development of new more efficient production processes, more attention is paid to production using the POX method. The partial oxidation chemical reaction is:

$$CH_4 + 1/2\, O_2 \leftrightarrow CO + 2H_2$$

Reforming processes that combine steam reforming and partial oxidation are the most efficient. These processes are called advanced processes, of which autothermal reforming (ATR) is one.

In the autothermal reforming process, both reaction one and three play an important role. The processes of POX and steam reforming are highly integrated. Both reactions take place in one reactor, which has similarities to the secondary reformer of the steam-reforming process. The reactions that take place are combinations of combustion and steam reforming. In the combustion zone, the reaction is:

$$CH_4 + 3/2\, O_2 \leftrightarrow CO + 2H_2O$$

This reaction is without CO_2 production because CO is the primary combustion product, which is converted to CO_2 by a slow secondary reaction. In the thermal and catalytic zones, reactions one and two occur to form H_2. The oxygen content of the oxidant in the reforming process depends on the application of the final product. For the production of NH_3, air is needed because it contains the N_2 necessary for the synthesis of ammonia.

Methanol reformers. Consider the methanol reforming reaction, which is being pursued as a source of hydrogen to fuel cells for mobile electricity generation on boats, for military hospitals, and for electric cars.

$$CH_3OH + H_2O \rightarrow 3H_2 + CO_2$$

This reaction can be modeled as occurring in two stages:

1. A nearly irreversible endothermic cracking reaction in which one mole of liquid methanol is converted into three moles of products:

$$CH_3OH \rightarrow 2H_2 + CO$$

2. Followed by the water–gas shift reaction:

$$CO + H_2O \rightarrow H_2 + CO_2$$

which is exothermic and equilibrium-limited.

For mobile applications, both reactions are typically performed in a plug–flow reactor using mixed copper–zinc oxide catalysts. Typically, this is followed by a purification step—a partial oxidation that removes unreacted CO. Without a membrane reactor, the heating and pressure requirements of this process are difficult, requiring a large reactor and significant heat transfer area.

If possible, the second reaction would be performed at high pressure and temperature to speed the reaction and for improved catalyst use. Because this reaction is very endothermic, the entrance temperature must be high, and heat must be provided along the reactor length. By contrast, low temperature and pressure are needed to drive the third reaction. Since this reaction is exothermic, heat must be removed either between this stage and the last, or along this part of the reactor.

This reformer uses partial combustion to reduce the CO content from 1 to 2% exiting the water–gas shift reactor to about 20 ppm. Hydrogen purity in this range is needed in order to use PEM fuel cells, and higher purity hydrogen would be better. It has been shown that, with a pure hydrogen source, the PEM fuel cell power density would be three times greater than with the 20 ppm of CO that this process produces. Thus, the fuel cell must be three times bigger than if pure hydrogen were available. Combustion catalysts also consume hydrogen, which reduces overall efficiency; they are used because catalysts the only low-pressure purification process available.

Thermocatalytic methane cracking. In this approach, methane is broken down into carbon and hydrogen in the presence of a catalyst at high temperature (850–1200 °C). The reaction is:

$$CH_4 \rightarrow \ + 2\,H_2 + C$$

This reaction is exothermic thus requiring a heat exchanger input of 10% of the feed gas. Florida Solar Energy Center researchers have studied this reforming technology. It is far from commercially available. The primary reasons are low efficiency of conversion and carbon fouling of the catalyst.

New reformer technologies

Researchers at the Pacific Northwest National Laboratory have developed a novel steam gasoline reformer with microchannels to reduce the size of automotive reformers. Plasma reformers use high-temperature thermal plasma (3000–1000 °C) to produce hydrogen from methane and hydrocarbon liquids. Thermal plasma accelerates the kinetics in reforming reactions even without a catalyst. The plasma is created by an electric arc. Reactant mixtures, such as methane or diesel fuel plus air and water, are introduced into the reactor and H_2 and other hydrocarbons are the products.

Drawbacks of reformers

Fuel processors also have drawbacks, including pollution and the overall low fuel efficiency.

Pollution

Although fuel processors can provide hydrogen gas to a fuel cell while producing much less pollution than an internal combustion engine, they still produce a significant amount of CO_2. Although this gas is not a regulated pollutant, it is suspected of contributing to global warming. If pure hydrogen is used in a fuel cell, the only byproduct is water in the form of steam. No CO_2 or any other gas is emitted. But because fuel cell powered cars that use fuel processors emit small amounts of regulated pollutants (e.g., carbon monoxide), they will not qualify as zero-emissions vehicles (ZEVs) under California's laws. Right now, the main technologies that qualify as ZEVs are the battery-powered electric car and the hydrogen-powered fuel cell car.

Instead of trying to improve fuel processors to the point where they will emit no regulated pollutants, some companies are working on novel ways to store or produce hydrogen on the vehicle. Ovonic, acquired by BASF in 2012, is developing a metal hydride storage device that absorbs hydrogen somewhat like a sponge absorbs water. This eliminates the need for high-pressure storage tanks and can increase the amount of hydrogen that can be stored on a vehicle.

Powerball Technologies wants to use little plastic balls full of sodium hydride, which produce hydrogen when opened and dropped into water. The byproduct of this reaction, liquid sodium hydroxide, is a commonly used industrial chemical.

Efficiency

Another downside of the fuel processor is that it decreases the overall efficiency of the fuel-cell car. The fuel processor uses heat and pressure to aid the reactions that split out the hydrogen. Depending on the types of fuel used, and the efficiency of the fuel cell and fuel processor, the efficiency improvement over conventional gasoline-powered cars can be fairly small.

4.2.9 Hydrogen storage

Storage of hydrogen is a problem. One gram of it occupies 22 liters of space when at standard atmospheric pressure. Hydrogen can be stored in many different ways. One, pressurization, uses standard but heavy containers. The weight and size constrain the usefulness of pressurization in vehicles. Hydrogen can be liquefied to increase its energy density. However, cryogenic-cooling techniques, which consume significant amounts of energy, are needed. Hydrogen can be stored in materials with chemical bonds. Metal hydride storage systems (MHSS) store hydrogen by chemically bonding with it. The metal can be any number of different elements and alloys.

The amount of hydrogen per volume is the typical figure of merit for a hydrogen-storage system. Compressed hydrogen at 200 bar has 0.99×10^{22} atoms per cubic centimeter. Liquid hydrogen, at $-253\,^\circ$C, has 4.2×10^{22} atoms per cubic centimeter. These values can be bested by a MHSS. Magnesium can store hydrogen with up to 6.5×10^{22} atoms per cubic centimeter. Table 4.5 lists the hydrogen densities obtained by storing hydrogen as a compressed gas, refrigerated liquid, and in several metal hydride materials. While the weight percentage of hydrogen is certainly not high in a metal hydride, the amount of hydrogen per unit volume is very good. Figure 4.41 shows a 650-liter hydrogen stored in a metal hydride.

The storage material alone is not the complete answer to efficient hydrogen storage. Both liquid and gas hydrogen storage require their own unique packaging, whether heavy high-pressure vessels or refrigerated Dewars. Likewise, the overall performance of the metal hydride materials depends on the construction and

Table 4.5 Hydrogen Storage Densities

Material	Hydrogen atoms ($\times 10^{22}$ per cubic cm)	Weight of material that is H_2
Hydrogen gas, 2850 psi	0.99	100%
Hydrogen liquid, $-253°C$	4.2	100%
Mgh_2	6.5	7.6%
Mg_2NiH_4	5.9	3.6%
$Fenih_2$	6.0	1.89%
$Lani_5h_6$	5.5	1.37%

FIGURE 4.41 A 650-liter hydrogen storage in metal hydride.

engineering of the container and hydrogen-handling system. Appropriate packaging of the metal hydride material can drastically improve the delivery characteristics of the hydrogen.

While the extraction of hydrogen gas from a pressure tank is simply a matter of using an appropriate gas regulator, using the gas pressure itself to transport the hydrogen. Metal hydride systems may use several methods to move the hydrogen from source, to storage, and to the fuel cell. Some metal hydrides respond to pressure. Increase the input hydrogen pressure by a few atmospheres and the hydrogen gas will migrate into the metal hydride. To retrieve the hydrogen simply release some pressure, or in some cases, draw a slight vacuum. Other metal hydride systems can be driven by heat. For example, slightly heating the metal hydride container will drive the hydrogen from the bonds where it is stored. Figure 4.42 shows various containers of metal hydride storage vessels.

Lightweight tankage is important for primary fuel cell powered vehicles that use onboard storage of hydrogen. In the past, energy-storage systems with extremely high specific energy (> 400 Wh/kg) have been designed that use lightweight tanks to contain the gases generated by reversible (unitized) regenerative fuel cells (URFCs). Today, the need for lightweight pressure vessels with state-of-the-art performance factors have been designed, and prototypes are being fabricated to meet the DOE goals (4000 Wh/kg, 12% hydrogen by weight, 700 Wh/liter, and \$20/kWh in high-volume production). These pressure vessels use technologies that are easily

FIGURE 4.42 Various sizes of metal hydride storage vessels.

adopted by industrial partners. URFCs are important to the efficient use of hydrogen as a transportation fuel and enabler of renewable energy.

URFC systems with lightweight pressure vessels, or hydrogen storage vessels, have been designed for automobiles. These energy-storage systems are expected to be cost competitive with primary fuel cell-powered vehicles operating on hydrogen/air with capacitors or batteries for power peaking and regenerative braking. These vehicles can be safely and rapidly refueled from high-pressure hydrogen sources, when available, to achieve driving ranges in excess of 380 miles. The employment of URFCs would save the consumer the entire capital cost of a home hydrogen-generation unit. That consumer would be able to electrically recharge at any available electrical source, instead of being tethered to a single electrolysis unit. UFRC-powered automobiles would still be able to rapidly refuel by direct hydrogen transfer when a hydrogen infrastructure becomes available.

Whether electrically refuelable, or not, a vehicle powered by compressed H_2 (at 5000 psi) is now the system to beat (as Defense Technology Institute has concluded). Such a vehicle offers: low weight while storing hydrogen in an acceptable volume, at an acceptable cost. Its other advantages include high system simplicity, excellent safety, and the potential for faster refills than its competitors, as well as expected support by a feasible H_2 infrastructure—in the both startup and mature phases.

4.3 COMPRESSED AIR ENERGY STORAGE

Compressed air energy storage (CAES) is a technology used to store energy by compressing air into a sealed location such as a cavern or a high-pressure tank.

4.3.1 Introduction

This process is carried out using an electric turbocompressor which pumps air into the cavern during off-peak hours; later on the high-pressure air inside the cavern is preheated and mixed with gas inside a turbine where, by the use of high- and low-pressure expanders, it will generate electricity during peak consumption hours. To generate electricity typically the compressed air is mixed with gas and burnt together the same way as in a conventional turbine plant. This method is more efficient than using only the compressed air to generate power.

CAES systems can be developed in natural geologic formations such as aquifers, solution-mined salt caverns, or constructed rock caverns. For mining purposes, caverns are about 60% more expensive since they are created by excavating solid rock formations. Aquifer storage is the least expensive alternative and therefore the most used. There is also another alternative called CAS or compressed air storage; in these types of systems, the air is stored into a fabricated high-pressure tank. However, the technology used for this alternative is far too costly considering that the storage space is small compared to CAES systems.

4.3.2 **Components of a CAES system**

A CAES system requires five different mechanisms: a motor/generator, an air compressor, a recuperator, control equipment center, and auxillary equipment. Figure 4.43 shows a schematic diagram of a CAES.

- *Motor/generator:* It employs clutches to provide for alternate engagement to the compressor or turbine trains. (1)
- *Air compressor:* This may require two or more stages, intercoolers and after coolers, to achieve economy of compression and reduce the moisture content. (1)
- *Recuperator:* In order to convert pressurized air into energy, first the air is preheated in the recuperator. This preheated air is mixed with a small portion of gas or oil and burnt in the combustor.
- *Equipment control center:* It is for operating the combustion turbine, compressor, and auxiliaries, and to regulate and control changeover from generation mode to storage mode. (1)
- *Auxiliary equipment:* The equipment needed consists of fuel storage and handling, and mechanical and electrical systems to support the various heat exchangers required. (1)

4.3.3 **Operation**

Conventional CAES systems store energy by driving large electric motors that pump compressed air into a mine. This process is done during off-peak energy demand when it is much less expensive. In addition, during the compression process the air is cooled down before injection in order to accommodate more air in the same space. The air will be pressurized to about 75 bars before entering the mine. Then, to convert this pressurized air into energy, first it is preheated in the recuperator. This

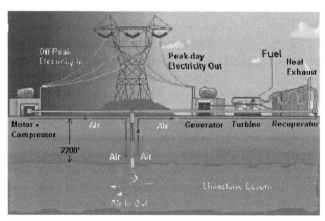

FIGURE 4.43 Schematic diagram of a CAES system.

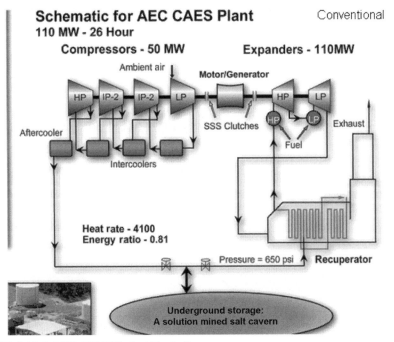

FIGURE 4.44 A CAES 110-MW plant block diagram.

preheated air is mixed with a small portion of gas or oil and burnt in the combustor. After that the hot gas from the combustor is expanded in the turbine to generate electricity. Figure 4.44 shows a 110-MW CAES plant block diagram.

4.3.4 Using CAES with a gas plant

According to . . . : "in a normal gas turbine, as much as two-thirds of the energy produced by the power turbine is consumed by the compressor. This means that a 300-MW plant actually produces 100 MW of net output while 200 MW is consumed by the compressor. If compression is carried out at a different time to when power generation is needed, that is, by CAES, the peak power that the gas turbine can generate is increased by this amount. Hence, additional power production capacity is not required. The turbine work that conventionally would drive the compressor in a combustion turbine is directed to the generator, increasing the output. In fact the power output from each turbine shaft will be nearly three times the capacity of a comparable simple cycle combustion turbine."

The reason why there isn't a compressed air turbine instead of a combustion turbine is because "the pressure and temperature conditions of the expanding air would be problematic. If the pressure was high enough to achieve a commercially meaningful output, then the temperature of the air at the discharge would likely be far lower than the materials and connections will be able to tolerate. This would result in embitterment of materials, seal failure, and perhaps leakage from differential expansion. While it might be possible to design a small air turbine that would work, this would probably require expensive, exotic materials and innovative engineering."(1)

4.3.5 Advantages and disadvantages of CAES systems

CAES systems can be used on very large scales. Unlike other systems considered large-scale, CAES is ready to be used with entire power plants. Apart from the hydro-pump, no other storage method has a capacity as high as CAES. Typical capacities for a CAES system are around 50 to 300 MW. The storage period is also the longest due to the fact that its losses are very small. A CAES system can be used to store energy for more than a year.

Fast startup is also an advantage of CAES. A CAES plant can provide a startup time of about 9 minutes for an emergency start, and about 12 minutes under normal conditions. By comparison, conventional combustion turbine peaking plants typically require 20 to 30 minutes for a normal startup. If a natural geological formation is used (rather than CAS), CAES has the advantage that it doesn't involved huge, costly installations. Moreover, the emission of greenhouse gases is substantially lower than in normal gas plants.

The main drawback of CAES is probably its reliance on a geological structure. There is actually not a lot of underground caverns around, which substantially limits the usability of this storage method. However, for locations where it is suitable, it can provide a viable option for storing energy in large quantities and for long times.[1] In summary, a CAES system: can be used in large scales, has a fast startup time, and is a low-cost installation when geological formations are used. The main drawback is its reliance on geological structures.

4.3.6 Worldwide CAES

So far there are only two CAES plants in operation in the world: the US 110-MW plant of Alabama Electric Corporation in McIntosh that was commissioned in 1991; and the 290-MW plant belonging to E.ON Kraftwerk in Germany, the first compressed air storage gas turbine power station in the world.

The McIntosh plant

The plant incorporated several improvements over the Huntorf one, including a waste heat-recovery system that reduces fuel usage by about 25%. The system has a rated capacity of 110 MW, from a roughly cylindrical salt cavern 300 m deep and 80 m in diameter (volume 5.32 million cubic meters); it supports pressures from 45 to 74 bars. The unit can supply power for 26 hours and takes between 9 (normal) and 13 minutes to start.

The Huntorf plant

This plant is located in northern Germany; it was commissioned in 1978 as the world's first CAES system. It was later extended with a 300,000-m³ natural gas cavern to supply the gas turbines with higher economic efficiency. (Figure 4.45). Table 4.6 shows the plant's basic design parameters.

The Huntorf plant consists of two caverns, though the total volume could have been realized with just one. This route was taken for several reasons. The advantages of splitting the volume between two caverns include redundancy during maintenance or cavern shutdown, and easier cavern refilling after drawing down its pressure to atmospheric pressure. The depth of the caverns was selected to be lower than 600 m. This was done to ensure the stability of the air for several months' storage, as well as to guarantee the specified maximum pressure of 100 bars.

The storage volume required is actually quite large (300,000 m³), about equivalent to forty 10-story buildings. This is why CAES can only use underground caverns; the cost and size of an above-ground facility of similar proportions would be staggering. Compression is done by an electrically driven air compressor and the system is charged over an 8-hour period, then delivering 300 MW for 2 hours of discharge. The experience after 20 years of operation is that the Huntorf plant runs reliably on a daily cycle and has successfully accumulated 7000 starts. The plant has reported high availability of 90% and a starting reliability of 99%.

4.3.7 Energy stored in a CAES

One type of reversible air compression and expansion is described by the isothermal process, where the heat of compression and expansion is removed or added to the

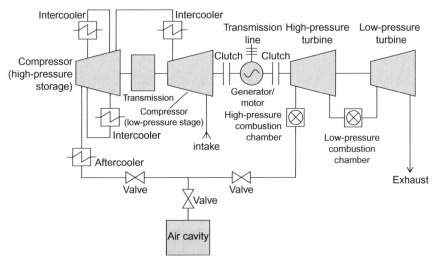

FIGURE 4.45 A diagram of the Huntorf plant.

Table 4.6 Basic Design Parameters of the Huntorf Plant

Output	
Turbine operation	290 MW (<3 hrs)
Compressor operation	60 MW (<12 hrs)
Air flow rates	
Turbine operation	417 kg/s
Compressor operation	108 kg/s
Air mass flow ratio in/out	1/4
Number of air caverns	2
Cavern location	
Top	650 m
Bottom	800 m
Maximum diameter	60 m
Well spacing	220 m
Cavern pressures	
Minimum permissible	1 bar
Minimum operational (exceptional)	20 bar
Minimum operational (regular)	43 bar
Maximum permissible and operational	70 bar
Maximum pressure reduction rate	15 bar/h

system at the same rate as it is produced. Compressing air heats it up and the heat must therefore be able to flow to the environment during compression for the temperature to remain constant. In practice, this is often not the case because to properly intercool a compressor requires a compact internal heat exchanger that is optimized for high heat transfer and low pressure drop.

Without an internal heat exchanger, isothermal compression can be approached at low flow rates, particularly for small systems. Small compressors have higher inherent heat exchange due to a higher ratio of surface area to volume. Nevertheless, it is useful to describe the limiting case of ideal isothermal compression of an ideal gas. The ideal gas law, for an isothermal process is:

$$PV = \eta\,RT = \text{constant} \tag{4.1}$$

By the definition of work, where A and B are the initial and final states of the system:

$$W_{A \to B} = \int_{V_A}^{V_B} P dV = \int_{V_A}^{V_B} \frac{nRT}{V} dV = nRT \int_{V_A}^{V_B} \frac{1}{V} dV = nRT(\ln V_B - \ln V_A)$$

$$= nRT \ln \frac{V_B}{V_A} = nRT \ln \frac{P_A}{P_B} = PV \ln \frac{P_A}{P_B} \tag{4.2}$$

where,

$P_A V_A = P_B V_B$
P = the absolute pressure
V = the volume of the vessel
n = the amount of substance of gas (the number of moles)
R = is the ideal gas constant
T = the absolute temperature
W = the energy stored or released

One mole of gas molecules at standard temperature and pressure (0 °C, 0.1 MPa), occupies 22.4 liters, and there are 1000 liters in 1 m³. So, there are about $\eta = 1000/22.4 = 44.6$ moles of gas molecules in 1 m³, and we get 101 kJ at 0 °C or 111 kJ at 25 °C per m³ of gas (at 0.1 MPa = aprox. atmospheric pressure).

It is very difficult to calculate the energy density for any unusually shaped volume. Therefore, if a piston, without friction, is assumed under an isobaric process, a simplified volume energy density can be calculated by:

$$E_{volume} = \frac{1}{V_0} \eta RT \int_{V_0}^{V} \frac{dV}{V} = \frac{1}{V_0} P_0 V_0 \ln \frac{V_0}{V} = P_0 \ln \frac{V_0}{V} \qquad (4.3)$$

where,

P_0 = the pressure
V_0 and V = the initial and final volumes

By assuming a starting volume of 1 m³ and a pressure of 2×10^5 Pa, if the gas is compressed to 0.4 m³ at constant temperature, the amount of stored energy is 1.8×10^J. This is a much higher energy density than that of magnetic or electric fields. In the case of CAES, the electrical storage efficiency cannot be compared directly to another power storage method due to the use of additional fuel. It is possible to calculate the net electrical efficiency of the CAES plant by subtracting the amount of energy generated by the fuel:

$$Storage\ Efficiency = \frac{\{P - F \times \eta\}}{E} \qquad (4.4)$$

where,

E = power consumed
P = the total power generated
F = the supplied fuel
η = the conventional generation efficiency (34%)

During experiments, sometimes the isothermal process is also considered to find out energy density instead of isobaric. In this process, the assumption is volume is kept constant and pressure of air is changed. This process is successful and applied in real practice. Research is ongoing in Korea with regard to a CAES plant. They are trying to follow the isobaric process and searching to improve human-made storage to get success.

EXAMPLE 4.1

For a 110-MW CAES plant, the air is stored in an underground cavern. It obeys the adiabatic process. If the temperature of the air in the cavern at 1 atm is 20 °C. Calculate raise in temperature if we increase the pressure to 100 atm. The volume of the tank is fixed.

Solution

For adiabatic process, pressure and temperature between gases is related as:

$$T_2 = T_1 \left(\frac{P_2}{P_1}\right)^{\frac{(n-1)}{n}}$$

For an ideal gas, $n=1$. Air is not an ideal gas, and it has $n=1.4 : T_1 = 20 + 273 = 293$ K.

$$T_2 = 293 \left(\frac{100}{1}\right)^{\frac{(1.4-1)}{1.4}} = 293 \times 100^{0.286} = 293 \times 3.73$$

$T_2 = 1093$ K, so temperature will rise to 720 °C.

EXAMPLE 4.2

The 50-MW CAES plant has a capacity to supply power for 36 hours. A cavern in salt rock that obeys the combine gas law. The temperature of the cavern is 800 °C. The before compressing air pressure of the air in the cavern is 700 kg/m². During off-peak hours, it is compressed to 3100 kg/m². The volume of the cavern is 350 m³. This system connects to a grid through a gas turbine. The conventional efficiency of the system with fuel is 42%. The energy supplied by fuel burning is about 14.4×10^6 MJ. Calculate:

a) Storage density of cavern
b) Storage efficiency of CAES plant

Solution

$$350 \, \text{m}^3 = 350 \times 10^6 \text{cm}^3$$
$$T = 800°C = 800 + 273 = 1073 \text{ K}$$

Energy density by volume:

$$E = VT \ln (P_1/P_0) = 350 \times 10^6 \times 1073 \times \ln (3100/700)$$
$$\approx 0.558 \times 10^{12} \text{ J}$$

Total energy generated capacity of the plant:

$$P = (50 \times 10^3 \text{ kW})(36 \text{ hours}) = 1.8 \times 10^6 \text{ kWh}$$
$$1 \text{ kWh} = 3.6 \times 10^6 \text{ J}$$

So,

$$P = (1.8 \times 10^6)(3.6 \times 10^6) = 6.48 \times 10^{12} \text{ J}$$
$$\text{Storage efficiency} = (P - F \times \eta)/E$$
$$= (6.48 \times 10^{12} - 14.4 \times 10^{12} \times 0.42)/(0.558 \times 10^{12})$$
$$= 0.7741 = 77.41\%$$

4.3.8 Future and planned construction of CAES systems

There are additional CAES plants built or planned. For example, Italy has operated a small 25-MWe CAES research facility based on aquifer storage. Research has been done in Israel to build a 3×100 MW CAES facility using hard rock aquifers. Similar projects have been started elsewhere to look into the possibilities of CAES systems.

Another plant currently under development is being designed by US Norton Energy Storage LLC. The site is a 700 m deep, 10,000,000-m^3 limestone mine in which they intend to compress air up to 100 bar before combusting it with natural gas. The first phase is expected to be between 200 and 480 MW and cost $50 to $480 million. Four more stages are planned to develop the site to a possible capacity of 2500 MW. (1)

4.4 FLYWHEEL STORAGE

Currently, we hear a lot of talk about the need to develop new sources of energy. As energy requirements increase, and the natural resources that presently help us produce energy become more and more scarce, it is necessary to look into obtaining energy in ways other than the traditional ones. However, since most of these new energy sources are not available at all times, it is also essential to develop ways of storing the energy when it actually can be produced. One of these ways is by flywheels. They might not be the most well-known method of storing energy, but they certainly have good capabilities, and if some economic problems can be overcome, they could be a very reliable energy storage means.

4.4.1 Principle of operation

Traditionally, a flywheel is a heavy metal wheel attached to a drive shaft, having most of its weight concentrated at the circumference. The wheel resists changes in speed and helps steady the rotation of the shaft where a power source, such as a piston engine, exerts an uneven torque on the shaft or where the load is intermittent. One of the big improvements James Watt made on the steam engine was to incorporate a flywheel to smooth the load.

The defining property of a flywheel is the spinning of a body on an axis. That single property can be as frivolous as the spinning of a top. The flywheel is both ancient and cutting-edge. It enables the steady rhythm of the porter's wheel and inspires the balance and orientation of the gyroscope. In marketer's definition, a flywheel is a practically lossless system for converting and storing electricity as kinetic energy of a magnetically levitated rotor assembly that spins in a vacuum and regenerating power as needed. It will enable ultrareliable, carefree electricity and no pollution; it is sustainable onsite alternative energy option. Basically, a flywheel is simply an electromechanical battery (EMB).

4.4.2 Basic components of a flywheel

Let us start by looking at the way flywheels are constructed. Many types of flywheels are in use and numerous other types are still being evaluated. It is not easy to predict which combination of shapes and materials will eventually prevail even if some appear to have limited development potential.

One useful parameter in assessing flywheel systems is the energy density of the rotor and the associated shape factor, K. Another parameter that expresses the ability of a certain flywheel to reach the design energy density is the velocity factor; this is a function the angular speed of the flywheel and the angular speed of a thin rim that has the same outer diameter and density and is equally stressed. An important issue in the construction of a flywheel is the choice of the material. This choice depends largely on the design requirements and on a variety of constraints. If there is no constraint on the peripheral velocity, the main feature is the specific strength of the material.

Rotor (the disk)

The energy density of a flywheel is proportional to the ratio between the maximum stress the material can withstand and its density. In this respect, modern high-strength composite materials can be considered the ideal choice. To design and build an EMB, recent advances in materials, such as high-strength fiber composites, particularly graphite, are taken advantage of. The strength of graphite fibers, now used in everything from tennis racquets to sailboat masts, has increased by a factor of 5 over the last two decades. These fibers play a central role in flywheel energy storage. Figure 4.46 shows the energy that can be stored in different materials that can be used in manufacturing of the flywheel.

The reason lies in the laws dictating how much kinetic energy can be stored in a rotating body. Any spinning rotor has an upper speed limit determined by the tensile strength of the material from which it is made. On the other hand, at a given rotation speed, the amount of kinetic energy stored is determined by the mass of the flywheel. This observation originally led to the intuitive notion that high-density

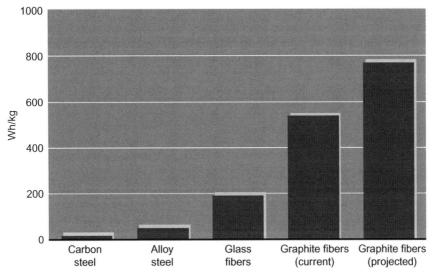

FIGURE 4.46 Flywheel energy stored in different materials.

materials, namely metals, are the materials of choice in flywheel rotors for energy storage.

A metal flywheel does indeed store more energy than an equivalent-size flywheel made of low-density material and rotating at the same speed. However, a low-density wheel can be spun up to a higher speed until it reaches the same internal tensile stresses as the metal one, where it stores the same amount of kinetic energy at a much lower weight. For example, lightweight graphite fiber is more than ten times more effective per unit mass for kinetic energy storage than steel.

Which modern fiber is optimum for an EMB depends on whether the designer wants maximum energy storage per unit mass (as in an application for a vehicle) or, for economic reasons, the designer requires maximum energy storage per unit cost (as in most stationary applications—e.g., load leveling for electric utilities). Vehicular uses call for graphite fibers even though these are more than ten times as expensive as the most cost-effective fiber for EMB stationary applications.

American physicist Post's emphasizes that using composite fibers has required the team to rethink the entire flywheel concept, which was based on metal flywheels. Because steel is an isotropic material, its strength against rupture is the same in every direction. Composites are typically anisotropic materials; that is, they are strong in the direction of their fibers but up to 100 times weaker in the other direction.

Laboratory flywheel designs use a basic geometry of a cylinder, with the fiber orientation that of a tight-wound spring; essentially it is perpendicular to the axis of the cylinder. In this way they achieve maximal strength in the outward centrifugal direction. The highest tip speeds attained by the rotor using the strongest available composite fibers vary from 1400 to 2000 m per second.

Housing (evacuated chamber)

We now take a closer look at some of the basic components of a flywheel. Some type of housing must be used to maintain it in a vacuum and to provide personal safety from contact, or, in the case of failure, from debris. A safety guard that prevents contact with the rotor is sufficient if the stresses are low enough to eliminate any danger of failure. Because the energy density is proportional to the maximum stress, total safety results in very low values of energy density. Higher stress levels introduce a remote possibility of a failure and make a containment structure necessary.

This practice is common in other fields of technology using rotating parts (e.g., aircraft gas turbines). These are surrounded by a containment ring that must be able to prevent the escape of any blade released in a failure. The need for operating flywheels in a vacuum usually occurs with medium- or high-energy density types and is necessary for two reasons: (1) to reduce the energy losses due to aerodynamic drag and (2) to avoid overheating of the rotor. Figure 4.47 shows a cross-section of a flywheel storage system.

Bearings

Another important component of flywheels is the suspension system. The bearings used in flywheel systems are often highly loaded, both by the static loads due to the weight of the rotor and in some cases by the forces due to the transmission units and by dynamic loads. As a combination of high loads and high speeds is often present, the design of the suspension system is often difficult. Further, a factor that complicates this problem is the need to minimize the drag torque of the bearings. Most

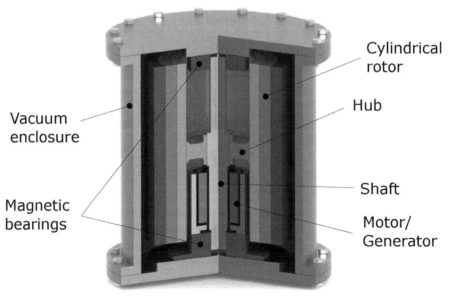

FIGURE 4.47 A cross-section of a flywheel storage system.

Source: Courtesy of Beacon Power

applications use rolling element bearings, but hydrostatic and hydrodynamic bearings have been used. Magnetic bearings, already used in a few selected applications, may be used in the future to improve performance of the system. Nevertheless, the choice of flywheel suspension systems depends on the type of transmission used.

Transmission system

The transmission system must allow a continuous variation in ratio and must operate efficiently under widely varying load conditions. The main types of transmissions are mechanical, electrical, and hydrostatic. The electrical transmission is the most common one. It uses motors and generators that can be placed within the vacuum housing. The only problem with this is the need of some sort of liquid.

The motor, which can be either brushless synchronous or induction, is fed from the mains through an inverter and a rectifier. The inverter gives a variable frequency input to the motor when the flywheel has to be charged. During discharge, the flywheel operates as the prime mover and the motor becomes a generator, its output being rectified and fed to a fixed-frequency inverter. In place of AC machines fed by inverters, it is possible to use DC machines controlled by choppers. The disadvantages of this solution are the greater size and cost of them; their somewhat lower efficiency; some speed limitations; and, above all, the inability to place the flywheel motor/generator directly within the vacuum housing due to the brushes.

4.4.3 Flywheel Operation

Power is supplied from the flywheel electronics module to drive the flywheel via the motor. When utility power is lost, the motor instantly acts as a generator and variable frequency, variable voltage power is supplied from the flywheel unit to its electronics module where it is converted to power compatible with the customer's load. This list follows the flywheel through its three modes of operation.

1. The flywheel charge mode
   - AC power is supplied to the flywheel electronics module (FEM)
   - Motor is spinning the flywheel at a high speed
   - FEM powers up the flywheel to full speed within a few hours
2. The flywheel float mode
   - Maintains full energy and speed with minimal AC power
3. The flywheel discharge mode
   - AC power to the motor is disrupted
   - It supplies mechanical power to the motor that becomes the generator
   - FEM provides uninterrupted power to the customer load
   - Speed decreases

4.4.4 Storing energy

The formula for energy storage in a flywheel is:

$$E = K \, MV^2 \qquad\qquad (4.5)$$

where,

> E=energy in joules
> K=a constant based on the shape of the flywheel
> M=mass of the flywheel in kilograms
> V=rotational velocity of the flywheel in m per second

Newtonian kinetic energy:

$$K = 0.5 \, J\omega^2 \tag{4.6}$$

where,

> ω=the body's angular velocity
> I=the body's moment of inertia

In the flywheels, we are especially interested in three types of I:

A solid cylinder is

$$I_z = \frac{1}{2}mr^2 \tag{4.7}$$

A thin-walled cylinder is

$$I = mr^2 \tag{4.8}$$

A thick-walled cylinder is

$$I = \frac{1}{2}m\left(r_1{}^2 + r_2{}^2\right) \tag{4.9}$$

Therefore, if the mass of the flywheel is doubled, the amount of energy is doubled. If you double the speed of rotation, you quadruple the amount of energy. The ratio of the energy stored over the mass is the energy density. Flywheels are often classified by their energy density. Although there is no general agreement on the meaning of low, medium, and high, the following definitions can be used:

> Low-energy density: lower than 10 Wh/kg (36 kJ/kg)
> Medium-energy density: between 10 and 25 Wh/kg (36 and 90 kJ/kg)
> High-energy density: higher than 25 Wh/kg (90 kJ/kg)

The laboratory's electromechanical battery (EMB), however, may be a better way to go or, at the very least, be an important piece in the evolving energy-storage infrastructure. Various devices are compared for peak power and specific energy. Batteries and flywheels produce roughly the same range of specific energy, but flywheels produce much more specific power than batteries.

EXAMPLE 4.3

Given a rotor mass of 5 kg, a diameter of 0.5 m, and spinning at 10,000 rpm, what is the amount of energy stored?

$$m = 5 \text{ kg}$$
$$d = 0.5 \text{ m}$$
$$v_1 = \pi d \left(\frac{10000 \text{ rpm}}{60 \text{ sec}} \right) = \pi(0.5 \text{ m}) \left(\frac{10000 \text{ rpm}}{60 \text{ sec}} \right) = 261.799 \text{ m/sec}$$

$$E_1 = \frac{1}{2} m(v)^2 = \frac{1}{2} (5 \text{ kg})(261.799 \text{ m/sec})$$
$$E_1 = 171346.791 \text{ kg·m/sec}$$
$$E_1 \approx 171 \text{ kW}$$

4.4.5 Installation

One of the examples is from the company called "beacon power." The Beacon Power Corporation's Flywheel Energy Storage System (FESS) is installed 18 inches below the ground surface and the flywheel electrical module (FEM) is installed above the ground. The installation process is performed in sequential steps that allows for a systematic installation. A hole is dug in suitable soil either with a backhoe or auger to a depth of 72" below grade and to a suitable width as indicated in Figure 4.48. This is done for safety reasons. A schematic diagram of the Beacon flywheel is shown in Figure 4.49.

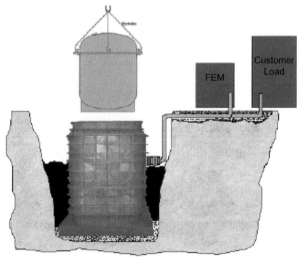

FIGURE 4.48 A hole for the burial of a Beacon flywheel.

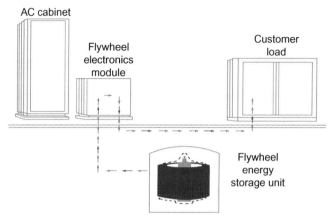

FIGURE 4.49 Schemaic diagram of the Beacon Power flywheel.

4.4.6 Advantages and disadvantages of flywheel storage

Advantages

The following list contains some of the advantages of flywheel storage:

High power density: This is most distinctive advantage. They can be charged at very high rates and can deliver very high powers. This feature of flywheels is exploited fully in some machines that need short bursts of high power. One of the promised applications is fast battery charging.

Nonpolluting: Another advantage of FESSs is their absence of pollution of all forms: chemical, thermal, or acoustical. The only instances at which flywheels can cause pollution are during their construction, in the case where some plastics are used.

High efficiency: The flywheel has a very high efficiency, even if the energy losses are dependent on the time during which the energy is stored. For short-time storage, the efficiency of a flywheel can be near 100%, which decreases for medium- or long-time storage. Advanced flywheels operating in a high vacuum suspended on magnetic bearings and can maintain a high efficiency for long periods; however, such systems have not yet been developed to an operational stage.

Long life: A 20-year operating life and thousands of charge/discharge cycles can be expected.

Climate/weather: Flywheels are unaffected by extreme conditions.

In addition, no maintenance is needed and flywheels can be installed underground to reduce visual clutter and increase available space.

Disadvantages

The following list contains some of the disadvantages of flywhell storage:

Safety: One disadvantage of flywheels is safety. It is difficult to build a structure that has to contain flying fragments caused by an explosion. Such a structure has a mass and a size much greater than the flywheel itself.

Noise: It is easy to overcome the noise from a flywheel, though this could be considered a disadvantage of them. Most noise is produced by the transmission. If properly designed, flywheels are very quiet, especially in vacuum operation.

High-speed operation leads to wear, vibration, and fatigue: This is directly linked to the general idea of kinetic energy storage which needs at least one fast-moving part with all the related problems of, wear, vibration, and fatigue.

4.4.7 Comparison with other energy-storage systems

A comparison with lead-acid batteries is shown in Tables 4.7 and 4.8. According to Table 4.8, flywheels are better for short-duration, high-power and high-cyclic applications and batteries are better for long-duration, low-cyclic applications. Overall, flywheels and batteries are complementary.

Table 4.7 A Comparison of Lead-Acid Battery and Flywheel Storage

	EMB	**Lead-acid battery**
Specific power	5-10 kW/kg	0.1-0.5 kW
Energy recovery	90-95%	60-70%
Specific energy	100 Wh/kg	30-35 Wh/kg
Service lifetime	>10 years	3-5 years
Self-discharge time	Weeks to months	Many variables (temperature, usage, etc.)
Hazardous chemicals	None	Lead, sulfuric acid

Table 4.8 Comparison of Flywheels and Batteries

Flywheels	**Batteries**
• Ideally suited for high power draw	• Operate best at low power draws
• Fast recharge, 10s of thousands charge /discharge cycles	• Slow recharge, 100s of charge/discharge cycles
• Low/mid energy (order of 1-25 kWh)	• Low/high energy (order of 1 K-1MWh)
• Accurate remote monitoring and predictable operation	• Monitoring less precise, lower certainty of operation
• Low to no maintenance	• Requires Maintenance
• Environmentally friendly—can bury in ground	• Lead-acid batteries require disposal procedures
• Little temperature sensitivity	• Narrow temperature operating range
• Emerging technology—cost potential	• Mature technology—very low cost

4.4.8 **Applications**

Flywheels are an excellent alternative to be used in supporting the power needs of telecommunications now and in the future. They are not new, but recent technological advances in composite materials, power electronics, vacuum technology, Internet communication, and efficient magnetic bearings have enabled a new generation of flywheels. The technology comes at the right time to support the distributed powering needs of the industry.

Flywheel energy storage can be more efficient than compressed air and thermal energy storage for certain applications. However, on a large scale, its energy storage is selected not for the technological feasibility of flywheels but rather for the economical one. Flywheel energy storage will be used on a large scale only if they show an economic advantage, at least in selected applications. This advantage must take into account the cost of the system itself, as well as the operating cost. For the time being, this has not been achieved, and most of the work is being done on the economical feasibility of flywheels.

Application of flywheel energy-storage systems to fixed machinery can be divided into the following four categories:

Low-energy applications: All applications in which flywheels store small quantities of energy in order to equalize the motion of a machine can be considered low-energy applications. Almost all flywheels used in industry fall into this category, as their main application is in mechanical presses or friction-welding machines.

Emergency devices: The use of flywheels can be of different types, ranging from simple ones connected to the shafts of cooling or lubrication pumps of machines, aimed at ensuring a safe slowing down in the case of power failure, to flywheels used to supply power to vital equipment for the entire period of a power failure.

Load levelers: An example of flywheels used as load levelers comes in the case when a highly variable electrical load has to be installed in a place away from the generating plant and power lines are not adequate. In this case, a load-leveling unit can be very cost effective.

Large-scale energy-storage devices: Finally, large flywheels have been proposed for the storage of large quantities of energy. Work is still being done on this issue in order to determine whether they are economically effective.

Flywheel energy storage can be more efficient than compressed air and thermal energy storage in certain applications. However, on a large scale, flywheel energy storage is selected only when other methods of storage are not suitable.

Flywheel energy storage will be used on a larger scale only if they show an economic advantage, at least for selected applications. This advantage must take into account the cost of the system itself, as well as the operating cost. For the time being, this has not been achieved, and most of the work is being done on the economical feasibility of flywheels. Only when the questions on this matter are erased can flywheels start to be widely used.

4.5 HYDROPOWER

Hydropower plants capture the energy of falling water to generate electricity. A turbine converts the kinetic energy of the water into mechanical energy. Then a generator converts the mechanical energy from the turbine into electrical energy.

4.5.1 Introduction

Renewable energy sources draw energy from the ambient environment as opposed to the consumption of mineral fuels (e.g., coal, oil, gas, uranium). The ultimate source of renewable energy available to man is the Sun. Its total radiant energy flux intercepted by the Earth is much greater than current capture-capacity renewable energy schemes. Despite the vast amounts of energy potentially available, collecting and utilizing that energy in an economical manner remains elusive.

The allure of renewable energy sources is quite promising and has fostered much interest. Renewable systems are independent of fuel supply and price variability, reducing economic risk factors. Renewable sources are more uniformly distributed geographically. Some forms of renewable schemes may be deployed in small units near consumer bases to eliminate costly transmission losses. Furthermore, the federal regulation of the US power system leads to great strides and incentives in the development and deployment of renewable sources.

Renewable energy sources are projected to gain future considerations in the domestic energy landscape as our understanding of the environmental affects of the burning of fossils fuels broadens. Currently, the greatest hindrance to wide-scale renewable deployment is the large initial capital requirements with respect to conventional power sources. Some forms of renewable energy systems produce only electrical power, which is of greater value than heat. Among these are hydro, wind, photovoltaic, tidal, and ocean power. Nonetheless, biomass, geothermal, and solar systems, which may be configured to produce both heat and electricity, are under similar research and development.

Renewable energy sources accounted for 13% of US electricity production in 1997 (Table 4.9). Worldwide renewable accounted for 21.6% of the total generated that year. The Greeks used the power of water wheels to grind wheat into flower more than 2000 years ago. In the 1700s, mechanical hydropower was used extensively for milling and pumping water. Prior to the invention of the steam engines, industrial processes were conducted using mechanical power generated by waterfalls flowing past water wheels. This meant industries had to be located in the vicinity of a river fall. The stream was then tapped upstream of the falls ran across a water wheel and discharged downstream of the waterfall.

The demand for mechanical power soon outgrew the inherent hydropower capacity that led to the advent of the steam engine as industry grew. Nonetheless hydropower accounted for 10.7% of US electricity production and 19% of the world's in 1990. Niagara Falls was the first of the domestic hydroelectric power complexes

Table 4.9 Energy Production of Renewables in 1997

Renewable resource	Electricity generation capacity potential (gigawatts)	Electricity generation potential (billion kilowatt-hours)	Renewable electricity generation as percent of 2012 electricity use
Wind			
Land-Based	10,955	32,784	809%
Offshore	4,223	16,976	419%
Subtotal	**15,178**	**49,760**	**1,227%**
Solar			
Photovoltaics	154,856	283,664	6,997%
Concentrating Solar Power	38,066	116,146	2,865%
Subtotal	**192,922**	**399,810**	**9,862%**
Bioenergy			
Subtotal	**62**	**488**	**12%**
Geothermal			
Hydrothermal	38	308	8%
Enhanced Geothermal Systems	3,976	31,345	773%
Subtotal	**4,014**	**31,653**	**781%**
Hydropower			
Existing Conventional	78	277	7%
New Conventional	60	259	6%
Subtotal	**138**	**536**	**13%**
Total	**212,314**	**482,247**	**11,896%**

Source: "U.S. Renewable Energy Technical Potentials: A GIS-Based Analysis", National Renewable Energy Laboratory. July 2012.

developed for large-scale commercial generation. The early hydroelectric plants were DC stations built to power arc and incandescent lighting during the 1880s. The invention of the electric motor brought rise to increased demands for electrical power. The post-World War I era marked the standardization of commercial hydroelectric complexes.

The Bureau of Reclamation played an instrumental role in the development of hydropower production because of its commitment to water resource management in the arid West. The waterfalls of the Reclamation dams held significant hydroelectric potential. Hydroelectric power generation has long been an integral part of Reclamation's operations, while it is actually a byproduct of water development. In the early days, newly created projects lacked many of the modern conveniences, one of these being electrical power. This made it desirable to take advantage of water's potential power source.

4.5.2 Major components and theory of operation of hydropower

Figure 4.50 shows the principles of operation of hydropower. Most conventional hydroelectric plants include the following major components:

- *Dam:* Most hydropower plants rely on a dam that holds back water, creating a large reservoir.
- *Intake:* Gates on the dam open and gravity pulls the water through the penstock, a pipeline that leads to the turbine. Water builds up pressure as it flows through this pipe.
- *Reservoir:* It is formed, in effect, as a storage device. The force of falling water couruses the against the turbine's blades causing the turbine to spin.
- *Water turbine:* This is much like a windmill, except the energy is provided by falling water. The turbine converts the kinetic energy of the water into mechanical energy.
- *Generator:* It is connected to the turbine. Its function is to convert the mechanical torque as a result of the falling water into electrical energy. Hydroelectric plant generators are very similar to those used at conventional power facilities.

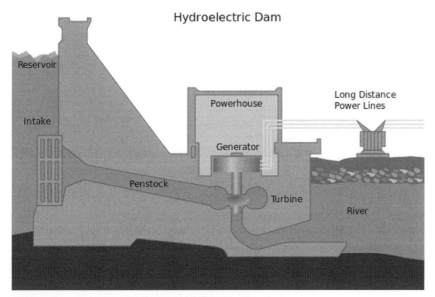

FIGURE 4.50 The principles of operation of hydropower.

- *Transmission lines:* High-voltage transmission lines transmit electricity from hydropower sites to residential and business consumers.

The amount of electricity a hydropower plant produces depends on two factors: (1) height of the water drop and (2) water flow rate.

The farther the water falls, the more power it imparts. Generally, the distance that the water falls depends on the size of the dam. The higher the dam, the farther the water falls and the more power that is produced. The amount of water available depends on the amount of water flowing down the river. Output power is directly proportional to the mass flow rate of the turbine inlet stream. Therefore, bigger rivers have more flowing water and can produce more energy. In general, the power output of hydropower, P, measured in watts is given as:

$$P = g \, QH \, \xi \tag{4.10}$$

where,

g = gravitational constant (9.8 m/s^2)
Q = flow rate, measured in liters/sec
H = water head, measured in meters
ξ = efficiency percentage of the turbine–generator pair

EXAMPLE 4.4

In a hydropower plant, the flow rate of the water through a penstock is 500 m^3/sec with an effective head of 100 m. How much power is generated if the turbine generator efficiency is $\eta = 75\%$?

$$P = \eta \, g \, Q \, h = 9.8.100.500.0.75 = 367.50 \text{ KW}$$

4.5.3 Types of hydropower turbines

At the heart of any hydroelectric power system is the water-turbine generator (Figure 4.51), composed of the generator and the proper type of turbine for the site. It is located in the powerhouse. The type of turbine specified is mostly determined by the head (vertical drop of the water flow) and the quantity of water available (measured in liters per second). The different types of turbines are (named after their inventers): the Pelton wheel, the Frances turbine, and the Kaplan turbine.

The Pelton wheel which resembles a water wheel with small bucket-like fins around its parameter, has many directional nozzles that direct the water to spin the turbine from all sides; it is for small amounts of water and a high head. The Frances turbine also resembles a water wheel but with spoke-like paddles sandwiched between two rims; it is enclosed by a central hub that has guide vanes around its inner parameter that direct the water to spin the turbine from all directions. This is one is used for medium head levels. Finally, the Kaplan turbine, which resembles a boat propeller, is flooded from the top and spins as the water drops by

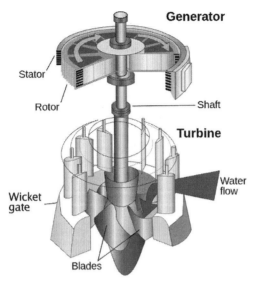

FIGURE 4.51 Cut-away drawing of a water-turbine generator.

its pitched blades; it is for low head levels because the pitch of the propellers can be adjusted to accommodate the flow. A drawing of all three is provided in Figure 4.52.

4.5.4 Hydro-pump storage principles

This is another form of hydroelectric plant that uses excess electrical system capacity, generally available at night, to pump water from one reservoir to another at a higher elevation. During periods of peak electrical demand, water from the higher reservoir is released through turbines to the lower reservoir, and electricity is produced. Although pumped storage plants are not net producers of electricity

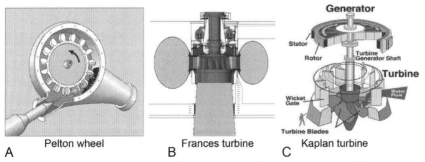

FIGURE 4.52 Three types of water turbine generators: (a) the Pelton wheel (b) the Frances turbine, and (c) the Kaplan turbine.

(it takes more electricity to pump water up), they are a valuable addition to electricity supply systems. Their value is in their ability to store electricity for use at a later time when demands are high. Storage is even more valuable if intermittent sources of electricity, such as photovoltaic or wind turbine systems, are hooked into the system.

Since electricity is the only commodity in existence that possesses the unique quality of needing to be consumed at the instant it is produced, as demand on power sources increased, so did methods of storing unused power. Like the storage plant, the pump storage plant uses the energy stored behind a dam to produce electricity as it is needed. But, when extra power is produced somewhere else, the surplus power is used to power the generator (now acting like a motor) to pump water through the turbine (now acting like a pump) from a reservoir at a lower elevation back behind the dam for use when demand is high. A photo of a pump storage plant located in The Snowy Mountains in Australia, and its two reservoirs and eight pump houses, can be seen in Figure 4.53.

This pump storage method proves to be exceptionally useful in cases where nuclear power is employed. Not only does the power from a nuclear plant need to be immediately consumed, but also due to the tremendous amount of heat produced by a nuclear plant, once started it has to produce power at full capacity on a continuous basis just to dissipate the heat. Twenty-four hours a day, at maximum capacity it would be impossible to operate a nuclear plant to match power demand without some form of storage, and hydroelectric pump storage seems to be a near perfect match for the job. Why you ask?

Nuclear energy only uses the power acquired from splitting an atom to produce heat. This heat is used to boil water into steam, which is used to turn a steam turbine

FIGURE 4.53 Photo of Tumut 3 Hydroelectric Power Station in The Snowy Mountains, NSW.

just like a conventional coal- or oil-fired plant. The extraordinary amount of heat produced by nuclear plants requires an extraordinary amount of water to cool the reactor. For this reason, nuclear plants are built near large sources of water, so automatically one component of a hydroelectric pump storage plant is already present near the nuclear plant.

One of the biggest advantages of pump storage is that hydroelectric power is considered to be 100% renewable; all the water is returned to the environment. That is not to imply that pump storage is 100% efficient, quite the contrary in fact. Whether the system is being used as a pump or a generator, the total efficiency is between 40 to 70% with the average system running about 50%. This means, in the case of the lost 40% in the previous example, that 50% of the unused energy from the 100-MW power station was stored as pumped water behind the dam and 50% of that was turned back into electricity, with a total of 10 of the 40 MW actually being converted into usable power. Why would we ever consider such an inefficient system to be a viable option?

First, energy management is only one consideration at hand. A properly planned hydroelectric system can provide irrigation, flood control, or even recreation facilities to a given region. Next resources: pump storage is rarely if ever built as a stand-alone station. They are usually built as standard storage sites that can pump up when extra power is available. Other advantages to pump storage are that they can be remotely operated and monitored, so there is a low manpower factor, as well as being relatively low maintenance.

In summation, pump storage is a clean, safe, reliable, renewable storage resource, with accountable and controllable affects on the environment. However, one of the disadvantages of hydropower is the silt build-up and the impedance to the movement of spawning fish.

4.6 SUPERCAPACITORS (ULTRACAPACITORS)

A capacitor is a device that stores energy in the electric field created between a pair of conductors on which equal but opposite electric charges have been placed. Energy can flow quickly in and out of capacitors with extremely high efficiency. A supercapacitor, which is also known as double-layer capacitor, ultracapacitor, electrochemical capacitor or pseudocapacitor is a new type of electrochemical energy-storage system. Accumulated energy is stored in the electric field of an electrochemical double layer.

4.6.1 Structure, characteristics, and operation of a supercapacitor

Conventional capacitors store energy electrostatically on two electrodes separated by a dielectric, while in an electrochemical double-layer capacitor (EDLC), or supercapacitor, charges accumulate at the boundary between electrode and electrolyte to form two charge layers with a separation of several angstroms. In addition, electrodes are constructed of high surface-area materials as carbon. Small distances between charged layers and large the electrode surface area enables EDLCs to achieve energy densities considerably higher than those offered by electrostatic capacitors.

Supercapacitors differ in the use of double-layered electrodes, which creates Faradic reactions through a surface charge–transfer. Current results from charging/discharging at the double layer. Most supercapacitors use activated carbon fiber for their electrodes providing a strong attraction to the ions in the electrolyte. Because the ions can be attracted to each other, a good electrolyte provides a good diffusion between each ion (Figure 4.54). Further research and development in supercapacitors has made them highly efficient through the use of advanced structured-carbon electrodes to optimize the charge and energy potential within a supercapacitor; this allows use in more advanced long-term energy-storage solutions such as electric vehicles.

One such company, Skeleton Technologies in Estonia, has made bold advances in supercapacitors by using nanostructured electrodes. They have the flexibility when developed to adjust the pores within the carbon electrodes for optimal interactions with the electrolytic material. By doing so, this allows higher ionic charges, thus containing more energy. In addition to the type of electrode that is used, the use of dual layers represents a capacitor at each electrode. In essence, you have two capacitors in series within a supercapacitor. Due to the dual-layered electrodes in a supercapacitor, the total capacitance can be regulated by adjusting the pore sizes in each electrode. The total capacitance is calculated as follows:

$$C_T = C_1 + C_2 \qquad (4.11)$$

This can be more readily seen in Figure 4.55.

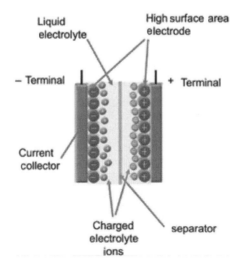

FIGURE 4.54 Ionic flow within a supercapacitor.

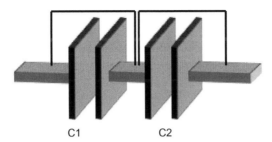

C1   C2

FIGURE 4.55 Series capacitor behavior in supercapacitor.

The energy stored in the supercapacitor can then be calculated as:

$$E = \tfrac{1}{2}\, C_T V^2 \qquad\qquad (4.12)$$

where, V is the voltage between C_1 and C_2.

EXAMPLE 4.5

A supercapacitor has C1 $=$ 50 F and C2 $=$ 50 F. The voltage across the supercapacitor is $V=100$ V; find the energy stored in capacitor E.

$$C\ total = C1 + C2 = 100\ F$$
$$E = 50 \times 10^5\ J$$

Given $E = 100$ J:

$$C1 = 50\ F\ C2 = 50\ F$$
$$C\ total = C1 + C2 = 100\ F$$
$$V = ?$$
$$V = (2E/C)^5 = (2 \times 100\ J/100\ F)^5 = 1.414\ V$$

In addition to double-layer capacitance, there is also observed a pseudocapacitance; this is a reversible redox, or ion absorption, Faradic reaction. A supercapacitor is capable of absorbing, storing, and discharging a very high current very quickly (Figure 4.56). A supercapacitor is composed of electrodes, separators and an electrolyte. The electrode is made of double-layered porous activated carbon. Increased surface area of the electrodes, leads to increased capacity. Since the energy stored in the capacitor is inversely proportional to the thickness of the dielectric, the electrode, separator, and electrolyte layers are made extremely thin, light, and tightly wrapped. This maximizes the energy-storage capacity of the supercapacitor.

When electrical voltage is applied across the electrodes, the current induces the formation of ions in the electrolyte material. The ions adhere to the surface of the electrodes (see Figure 4.56). The ions do not migrate between electrodes. When an

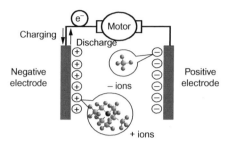

FIGURE 4.56 Physical charge absorption.

electrical load is applied across the electrodes, the supercapacitor quickly discharges when the ions lose their charge and get released back into the electrolyte. The charging and discharging process is completely reversible in the supercapacitor. The electrodes and electrolytes do not lose their structure and composition over time. Thus, the supercapacitor lasts a very long time and is capable of delivering more than 100,000 cycles. The voltage varies linearly with the state of charge (i.e., the voltage across the supercapacitors drops linearly as it discharges through the load).

Electrode materials
The electrode material for a supercapacitor must exhibit very high specific surface area because double-layer charge storage is a surface process. There are three types of electrode materials suitable for the supercapacitors: high surface area activated carbons, conducting polymers, and metal–oxides.

Carbons
In the majority of commercial applications, carbon electrodes are used. They present advantages in low cost, availability, and wide commercial acceptance. These electrodes can come in a variety of forms such as foam, fibers, and nanotubes. In addition, Faradic reactions are common to carbon electrodes. Treatment of activated carbons can make use of pseudocapacitance and further improve device performance.

Conducting polymers
In conducting polymers, charge storage and release is achieved using a redox reaction. High specific capacitance is attributed to the fact that charging takes place though the bulk volume of the film. Although conducting polymers have been reported to have high-power and high-energy densities, there use is still limited due to the swelling and shrinking effect.

Metal–oxide electrodes
Metal–oxide electrodes possess high specific capacitance and low resistance but are an expensive alternative to carbon. Extensive research has been done by the military; for example, the US Army Research Lab assembled prototype cells with an energy

density of 8.5 Wh/kg and a power density of 6 kW/kg. Metal–oxide electrodes can only be used with aqueous electrolytes, thus limiting the achievable cell voltage. Gains in power density from lower resistance are therefore often offset by losses due to the lower operating voltage.

Electrolyte materials

Electrolyte material can be either aqueous or organic. Electrolytes breaks down voltage and limits the cell voltage of a supercapacitor.

Aqueous electrolytes

Bases or acids KON, H_2SO_4) both are suitable for carbon or metal–oxide electrodes. Even though aqueous electrolytes are less expensive, easier to purify, and have a lower resistance, they typically limit cell voltage to 1 V, whereby the maximum achievable power is also limited.

There is a special interest in the aqueous electrolytes compared to the organic ones; that is, to enable the existence of an additional pseudocapacitance on the redox reaction of the hydrated functional groups at the surface of the materials. Utilizing these phenomena, for example, can double the value of the capacitance on a carbon material in an aqueous electrolyte compared to the same material in an organic electrolyte.

Organic electrolytes

Currently, most of the electrochemical double-layer capacitors use organic electrolytes. They have higher internal resistance but also they allow higher cell voltage of 2 V and higher. Higher internal resistance usually does not limit the power performance of the device due to the offset by the gain in higher cell voltage. According to some sources, acetonitrile is used in 90% of the large-size organic electrolyte capacitors. Acetonitrile displays a sufficient conductivity to enable an almost unchanged cell performance at $-30\,°C$. The only drawback of this electrolyte is its toxicity.

Separator

The separator has dual function in EDLC: to prevent the electrical contact between the two electrodes but to allow ion flow to ensure the charge transfer. With organic electrolytes polymer or paper separators can be used and with aqueous electrolytes ceramic or glass fiber separators are often used.

4.6.2 Advantages and disadvantages of supercapacitors

Supercapacitors can be used in wide variety of applications such as electric vehicles, fast battery charging, and uninterruptible power supplies. The advantages and disadvantages of supercapacitors are a result of their structure and characteristics.

Advantages
- *High power density (W/kg):* Ability to deliver and absorb high power and current quickly, which is beneficial for acceleration, regenerative braking,

and bridging the gap in UPS. Due to their rapid charging and discharging capabilities, supercapacitors are used in fast battery charging applications.

- *Low resistance:* Ability to deliver high current without a lot of heat loss in the capacitor
- *High efficiency:* 90% or more
- *Quick response time:* This is beneficial in bridging the gap in UOS and smoothening spikes in current demand
- No maintenance
- Long lasting: more than 100,000 cycles
- Robust solid state device

Disadvantages

- *Incomplete capacity utilization:* Supercapacitors can sustain 100% depth of discharge (DOD) but since the voltage reduces linearly with discharge, only a limited portion of the capacity is discharged before reaching the cut-off voltage of the load circuit.
- *Low energy density (Wh/kg):* The amount of energy stored by a supercapacitor is small compared to the needs of applications such as electric vehicles. Thus, they are used in conjunction with other energy-storage and delivery devices (e.g., batteries and fuel cells).
- *Low voltage per cell:* Since the voltage per cell is low, a large number of supercapacitors are needed in series to service applications such as electric vehicles. This also increases the cost.
- *High self-discharge rate:* More than 50% per month
- Requires voltage correction circuits to compensate for the drop in voltage during discharge
- Sensitive to temperature

4.6.3 Differences between supercapacitors and other energy-storage methods

A battery is structurally similar to a supercapacitor but the ions migrate from anode to cathode or vice versa. Since the ions physically migrate, batteries have a limited rate of charge or discharge. On the other hand, batteries have a higher energy density compared to supercapacitors. In a battery, the composition and structure of the electrodes and electrolyte changes with the state of charge, and they do not get fully restored at the end of a cycle. Therefore, battery life is limited while supercapacitors can last 100,000 cycles.

Like supercapacitors, flywheels have very high power density but low energy density. Thus, flywheels are suitable for fast charging and discharging application. Flywheels store energy in the form of high-speed spinning wheels and are very heavy and complicated compared to supercapacitors. Table 4.10 summarizes the differences between some commonly used energy-storage and delivery devices.

Table 4.10 A Comparison of Energy-Storage Technologies

	Supercapacitor	Battery	Fuel cell	Flywheel
Method of operation	Charge stored as ions on electrode surface. Physical absorption of charge	Ionization and migration of ions between electrodes through the electrolyte and separator. Electrons migrate through the load circuit.	Oxygen and stored or reformed hydrogen are ionized in the presence of a catalyst. The ions migrate, react, and form water. Electrons migrate through the load circuit.	Energy is stored in the form of a high-speed rotating wheel (up to 1666 rev/sec). Energy is converted to mechanical or electrical.
Power density (w/kg)	1–10 kw/kg	Up to 400	Various	1–200 kw/kg
Energy density (wh/kg)	1–5 kWh/kg	Up to 650 wh/kg	Various	10–900 wh/kg
Life (cycles)	100,000	Up to 1000	Various	10,000
Operating temperature	Limited range	Limited range	Predefined limited range	Wide range
Maintenance	Low	Various	Various	Low
Weight	Low	Various	Various	High
Efficiency	High	Various	High	High
Cell voltage	Low	Various	Various	N/a
Self-discharge rate	50%/month	Up to 30%/month	N/a	Low
Fast charge capability	Yes	Various	Yes (fill hydrogen in tanks)	Yes

4.6.4 **Applications of supercapacitors**

Electric vehicles

Supercapacitors, when used as key components of an electric vehicle (EV), greatly improve their performance, efficiency, and range, and they lower their purchase and operating costs.

Performance and acceleration

Supercapacitors have a very high power density of 1500 W/kg and above. This property is very useful during acceleration of the electric vehicle from standstill or at high speeds or for fast climbing. The supercapacitor is placed in parallel with the battery in battery-powered electric vehicles (BPEV) or in parallel with the fuel cell in fuel cell-powered electric vehicles (FCEV). Since the supercapacitor voltage is low per cell, a number of supercapacitors are setup in series to match the voltage of the battery or fuel cell.

During acceleration, when the throttle is depressed, the electric motor draws a high current from the power source. The supercapacitor delivers the high current using stored energy. The battery and fuel cell do not need to have a very high power rating. The peak power output of the entire power source is greatly increased due to the supercapacitor. The battery or fuel cell charges the supercapacitor when the vehicle is coasting or when the output of the battery or fuel cell is more than the needs of the electric motor.

Regenerative braking

When the electric vehicle is decelerating, the electric motor acts like a generator and pushes a current back to the power source. A supercapacitor setup in parallel with the fuel cell can quickly absorb and store this energy to be used later for acceleration. In this way, the supercapacitor is able to trap and store the kinetic energy of the car that would otherwise be lost as heat through the brakes. Thus, regenerative braking improves the electric vehicle's fuel efficiency and range.

Sizing of the battery or fuel cell

The supercapacitor acts as a peak load-buffering device that smoothes out the peaks and valleys in the power demand cycle. Therefore the size and power rating of the battery or fuel cell need not be very high. This reduces the overall cost of the electric vehicle. Due to its meager energy-storage capabilities (low specific energy 3.9 Wh/kg), the supercapacitor cannot fully replace the battery or fuel cell. It delivers the peak power while the battery and fuel cell deliver the range of the electric vehicle.

Fast charging of batteries

Because supercapacitors have a very high specific power rating, they are able to deliver a large amount of charge very quickly. Moreover, they are robust and have a very long life. Thus, supercapacitors are very effective in commercial and residential battery fast-charging devices. Such equipment is expensive and usually is used in off-board or mobile battery chargers. The off-board equipment and its cost can be shared by multiple EVs to charge their onboard batteries.

Fast-charging equipment draws AC power provided by the electric utility. The AC voltage is converted to DC using a rectifier and fed into a bank of low-voltage supercapacitors. Since supercapacitors have a low per cell voltage, a large number of supercapacitors are placed in series. A DC/DC converter then steps up the voltage of the low-voltage supercapacitor bank and feeds it into a high-voltage supercapacitor bank through an electrical switch. Batteries in electric vehicles are charged using a meter and the bank of high-voltage supercapacitors. A controller controls the operation of the meter, supercapacitors, switch, and converter.

In the case of mobile chargers, the bank of supercapacitors can be charged at the service station and then transported to any location along the highway to assist disabled electric vehicles. Since supercapacitors are light and compact, they are suitable in such mobile fast-charging applications. Such sophisticated off-board battery chargers eliminate the need for onboard battery chargers, thus making the electric vehicle lighter, less complicated, and lower cost.

UPS

Short-term bridging in uninterruptible power supply (UPS) systems is needed when there is power outage; the backup systems need some time to come online. Supercapacitors are used to deliver the power quickly during the time gap between the power outage and the startup of the backup systems. Backup power is usually supplied by diesel generators or fuel cells.

Capacitive starting of heavy diesel engines

Heavy diesel engines in trucks and locomotives that are stopped need a large amount of power to start. Since supercapacitors have a very high power density, they are used to deliver the high current needed to start such heavy engines.

Unmanned storage for solar or wecs

Supercapacitors can be used to store the energy of the renewable energy sources such as wind energy, photovoltaic, and solar energy.

4.6.5 Other supercapacitor applications

Other usages for supercapacitors include the following:

- High-powered lasers
- CMOS logic circuits
- VCR circuits
- CD players and high-fidelity stereos
- Cameras

- Computers
- Clock/radios
- Telephones and cellular devices
- Fire alarms
- Office equipment
- Actuators

4.7 SUPERCONDUCTING MAGNETIC ENERGY STORAGE

In times when technology is growing at a fast pace, the need for energy storage grows at the same rate. The discovery of energy sources in past times has revolutionized our present way of life. The high demand for energy is creating a necessity for those traditional energy sources to be optimized. An upcoming technology for storing energy is superconductors in the form of Superconducting Magnetic Energy Storage (SMES) systems.

4.7.1 Introduction

Since the current main source of energy is natural resources, as time goes by we may find a lack of traditional natural source. The supply and demand of energy may become a problem in the near future. To keep up with these demand scientists, physicists, chemists, and engineers are often researching and developing renewable alternative energy sources. Nevertheless, energy from the upcoming energy sources is not obtainable always; therefore, it is also important that ways of storing the produced energy be developed.

SMES systems store energy from a source in cryogenically cooled superconductive coils with no losses, which lets current flow in a "forever" loop. The SMES system is often referred to as the magnetic battery. Superconductors behave more like a perfect conductor; this characteristic allows the superconductor to carry direct current with no losses because there is no resistive heat to dissipate the energy.

When induced into the superconductor, the direct current finds itself in an infinite loop. Alternative current can also be induced by the superconductor but with time you will find a small loss of energy. Superconductors achieve perfect conductivity by having zero resistance. The zero-resistance level is achieved by maintaining a temperature below a critical temperature. Even then, to have perfect conductivity, the level of current density and magnetic field may not rise above a critical point.

Superconductivity can be defined as the loss of resistance in a material as current sweeps across it. It was first proven in 1911 by a Dutch physicist. Many discoveries and technological applications make it promising this Nobel Prize field. However, it is only been in recent times, when higher level of superconductivity has been reached, that has allowed for promising applications. Today, superconductivity applications are diversified in use, including medicine, theoretical and experimental science, the military, transportation, power production, energy storage, electronics, and many more.

4.7.2 History of superconductivity

Dutch physicist, Heike Kammerlingh Onnes in 1911, researched the electrical properties of metals in low temperatures. In one of his experiments, Onnes drove a current flow through mercury as he lowered the temperature of the material. He discovered that around $4\,°K$ the resistance of the mercury to electricity was gone. He used liquid helium to reach the low temperatures. Two years later, Onnes was awarded a Nobel Prize for his work and discovery—he called superconductivity. If liquid helium is used to cool a superconductor, it is called a low temperature superconductor (LTS). During Onnes's lifetime he kept researching superconductors and also noted that electrical currents, magnetic fields, and temperature influence superconductors.

Years later, in 1933 Walther Meissner and Robert Ochsenfeld discovered that superconductors not only have zero resistance but also have special magnetic field characteristics. They discovered that at low temperatures superconductors repel magnetic fields. The induced current that flows inside the material creates a magnetic field that is a replica of the magnetic field that was supposed to penetrate. Thus, a magnet can be swept up by the repulsed magnetic field. The occurrence was named the Meissner effect. The Meissner effect is affective only if the magnetic field is small enough; a large magnetic field will penetrate the material.

Later in 1957, the BCS theory was develop by American physicists John Bardeen, Lean Neil Cooper, and J. R. Schrieffer. The theory describes the superconductivity phenomenon in quantum physics as paired electrons, also known as Cooper pairs, which carry electrical current together. They stated that no one single electron has that characteristic. The progress was then developed more by the confirmation of the Josephson effect. Brian Josephson predicted that an electron will travel through a thin insulating layer in between two superconductor materials. Yet another Nobel Prize-wining characteristic of superconductors.

The next big development for the technology was the contribution of researchers at the IBM Zurich Research Lab. in Switzerland. Alex Müller and Georg Bednorz were making studies with ceramic material and how it behaves under low temperatures. They discovered that around 35 K the ceramic insulator began to acquire superconductance characteristics. Another Nobel Prize- winning performance for the technology, which led other researchers to experiment with more ceramics and compounds. A year later a ceramic was found to be superconductive at an all-time high of 90 K.

The discovery of the higher-temperature compounds allowed researchers to use liquid nitrogen as a coolant. If liquid nitrogen is used to cool a superconductor, it is called a high-temperature superconductor (HTS), even though temperature is still relatively low. The standing world record for a temperature in superconductance is 138 K. The SMES technology is now driven by research and development of materials that would be of much higher temperature.

4.7.3 Components and operation of SMES

SMES system components are the following: superconducting coil, power conditioning system, cryogenically refrigerator, and a vessel with a cryostat/vacuum for a low-temperature coil (Figure 4.57). The operation of a SMES system is simple. A source voltage is fed into the superconducting coil. Whenever the voltage from the source is AC voltage, it must be converted to DC. Most renewable energy application sources do supply DC voltage Because of the power transfer from the source to the coil.

The design of the device will limit the maximum energy stored. The power-conditioning system, or power electronic device, makes the decisions on when to feed power for charging purposes, or when to take energy from the SMES to provide power. When energy is needed, currents induced on the coils are superimposed on the load as specified by the control system. If needed, a DC-to-AC current conversion can be done.

During energy storage in the superconductor coil, the cryogenic refrigerator keeps the coil at low temperature for the superconductivity characteristic needed to have no resistive loss. No resistive loss means the current stored will flow in the coil until the moment of utilization. The power-conditioning or control device uses solid-state electronics. The wiring diagram of an SEMS system is shown in Figure 4.58.

4.7.4 Energy stored in the SMES

The magnetic energy stored by a coil carrying a current is given by one half of the inductance of the coil times the square of the current

$$E = .5 \times L \times I^2 \qquad (4.13)$$

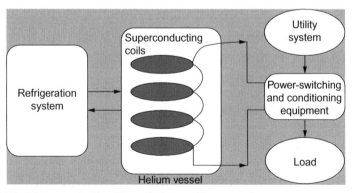

FIGURE 4.57 A Superconducting Magnetic Energy Storage system.

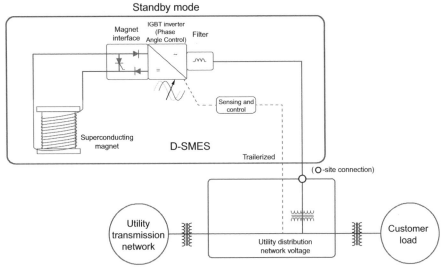

FIGURE 4.58 A wiring diagram of a SEMS system.

where,

E = energy measured in joules, J
L = inductance measured in henries, H
I = current measured in amperes, amp

EXAMPLE 4.6
A laboratory uses a SMES current, I, of 1000 amps. The average power used is
200 kW and the maximum is 800 kW. The magnetic field is 4.5 T and the inductivity is 4.2 H. How much energy should be stored by the coil in the SMES system
to ensure the power quality of the laboratory plant?
    HUsing the equation (E = .5 * L * I2), we plug in the values for it:

- Inductance = 4.2 H
- Current = 1000 A

$$E = (.5) \times (4.2) \times (1000)^2$$
$$E = 2,100,000 = 2.1 \text{ MJ}$$

4.7.5 Applications of superconducting magnetic energy

The following are some of the applications of a superconducting magnetic energy
system:

- Storage for renewable energy systems such as photovoltaic and wind energy
  conversion systems.

- SMES systems offer utilities a cost effective way to maintain grid-level voltage stability and protect from momentary voltage drops. SMES systems increase transfer capacity and protect utility grids from the destabilizing effects of short-term events such as voltage dips caused by lightning strikes and downed poles, sudden changes in customer demand levels, and switching operations.
- *Magnetically levitated train:* Research and development of a superconducting magnetic levitation (maglev) railroad have been in the experimental stage since 1977 in Japan. In 1990, the so-called Yamanashi Maglev Test Line was introduced and began operation by some Japanese companies: Central Japan Railway Company, the Railway Technical Research Institute, and Japan Railway Construction Public Corporation. The project was run with contributions of two superconducting magnets provided by Hitachi (a Japanese company). The train reached a world record of 552 km/h at the end of the last decade (Figure 4.59).
- *Magnetic Resonance Imaging (MRI)*: A medical diagnostic technology that outputs images of the inner human body. Most major hospital now have an MRI machine. The machines are large in size and cost; this due to the use of a low-temperature superconductor and the amount of space and material needed to keep the liquid helium from evaporation.
- *Wires:* Still an up-and-coming application, having no resistance in wire will allow current to travel without any losses to create a higher efficiency in energy and power transfer. The HTS wire can carry a current that is exponentially bigger then that of regular copper, with a relatively smaller diameter. (See Figure 4.60.)
- *Flywheel bearing:* Flywheel technology uses superconductivity to levitate the flywheel on superconductor bearings so it can store kinetic energy without losses, with the zero-resistance characteristics of HTS. A superconductor behaves like a spring with large damping due to hysteresis.
- *SQUID:* That is, the superconducting quantum interference device detects a magnetic field over the surface by detecting the magnetic flux displacement and

FIGURE 4.59 A magnetically levitated train.

FIGURE 4.60 Future application of SEMS in overhead lines.

flow rate of energy. It is a device that uses the Josephson junction—two superconductors separated by an insulating layer. SQUIDs are use by physicists in applications to gravitational waves. They have also been used by geologists to find oil and deposits.

4.7.6 **Advantages, disadvantages, and the future of SEMS systems**

The following lists contain some summary information.

Advantages

- *High power*: Power ranges from 1–10 MW to systems that carry 10–100bMW
- *Low discharge time*: Releases large amount of energy in small fraction of a cycle
- *High efficiency:* Due to no resistive heat loss, the efficiency of a SMES is relatively high relative to any other storage system. Since the energy is stored as a magnetic field there has to be no conversion to any other technology, only an electrical-to-electrical DC/AC converter if AC voltage is needed.
- *Environmentally safe*: SMES sytems are environmentally safe because no chemical reaction is produced by the technology. In addition, no toxins are produced in the process.

Disadvantages

- *Size:* to achieve high energy storage you need long distance of looped superconductive coil.
- *Electromagnetic interference:* Large installations will require a relatively large buffer zone because of relatively high magnetic fields that would be created, which would dwarf the Earth's magnetic field.
- *Cooling*: It hasn't been until the recent development of superconductivity that liquid nitrogen has been able to be used as a coolant. In past times, liquid helium was the only coolant and that gave temperature problems to achieve superconductivity.

- *Expensive:* Having a coolant a system that comes along with each individual unit makes budget a difficult to overcome characteristic of SMES systems.

Future
- *Utility grid wires*: As HTS systems are developed, the zero-resistance characteristics will revolutionize the worldwide applications of grid wires, and high current flow will be available with no loss across the wire, no matter the distance.
- *HTS MRI:* The LTS magnets in most MRI machines will be changed into HTS allowing for size and cost to be minimized and allow for easier accessibility for smaller, lower budget clinics.
- *Room temperature superconductors*: The development of material that would not require complicated coolant will allow for SMES systems to be smaller size and less costly.

PROBLEMS

4.1   A creek has the flow rate of 1000 m³/sec. It flows through a penstock with an effective head of 10 m. How much power is generated by a hydropower plant operating with a turbine–generator efficiency of $\eta = 76\%$.

4.2   In a dual-layered electrode supercapacitor, the capacitor value is summed up to be 600 μF. The rated voltage for it is 330 V. What is the amount of energy stored in the supercapacitor?

4.3   A flywheel is in the shape of a sphere. Given that the rotational velocity of it is 30×10^3 rpm with a 140×10^3 g mass, calculate the amount of kinetic energy availible, if $K = 2.5$ for a sphere shape.

4.4   For a hydropower station, the flow-rate of water is recorded at 250 m³/s. The efficiency of the turbine generator is 62% with the water head measured at 75 m. How much power is generated?

4.5   A flywheel energy-storage system has the following parameters: M = 135 kg, operating speed range of 15,500 rpm, diameter of 24 in. (610 mm), and the rotor is a thin-walled cylinder. Calculate the stored energy.

4.6   A supercapacitor stores E = 100 J. If $C_1 = 50$ F and $C_2 = 50$ F, what must be the voltage across the supercapacitor?

4.7   A laboratory uses a SMES current, I, of 100 amps. Inductivity is 2.2 H. How much energy should be stored by the coil in the SMES system?

4.8   For the Three Gorges Dam hydroelectric power plant in China, which is the largest in the world, the dam is up to 175 m high, has an average of 47,300 m³/sec, and the turbine–generator efficiency is 95%. Calculate the power output.

4.9 If there is 50 moles of gasoline (n) in a vessel that has a volume of 1 m^3 then what is the energy stored in this device if the absolute pressure, P, is 10 kJ? Assume that the final pressure, Pb, and the initial pressure, Pa, have a 10% difference.

4.10 A flywheel desk has a radius of 1 m, a weight of 10 kg running at a speed of 36,000 rpm. Calculate the energy, in kWh, that can be stored in this flywheel.

References

[1] *http://en.wikipedia.org/wiki/Nickel-cadmium_battery.
[2] *www.utc.edu/Research/CETE/electric.php.
[3] *http://thefraserdomain.typepad.com/energy/2006/01/sodiumsulfur_na.html.
[4] *http:/newenergyandfuel/com/2008/03/05/the-biggest-batteries-ever-answers-in-part-for-storing-wind-energy/.
[5] *http://energystorage.org/tech/photo_vrb1.htm.
[6] *www.electricitystorage.org/tech/technologies_technologies_psb.htm.
[7] http://fedgeno.com/chronos/egypt.html.
[8] http://energystoragedemo.epri.com/cec/zbb/tech_desc.asp.
[9] Biography of Sir William Grove, www.corrosion-doctors.org/Biographies/GroveBio.htm.
[10] www.xcelenergy.com/SiteCollectionDocuments/docs/AppendixE.pdf.
[11] http://en.wikipedia.org/wiki/Fuel_cell.
[12] *http://americanhistory.si.edu/fuelcells/basics.htm.
[13] *https://www.llnl.gov/str/Mitlit.html.
[14] http://americanhistory.si.edu/fuelcells/images/sox4.jpg.
[15] www.fuelcells.org/basics/apps.html
     *www1.eere.energy.gov/hydrogenand**fuelcells**/pdfs/specialty_vehicles.pdf.
[16] *www.nfcrc.uci.edu/2/FUEL_CELL_INFORMATION/FCexplained/FC_Comp.aspx.
[17] *Patil PG. Fundamentals of fuel cells. Washington, DC: U.S. Department of Energy.
[18] *Hydrogen the fuel for the future. U.S Department of Energy (DOE) by the National Renewable Energy Laboratory, a DOE National Laboratory; March 1995.
[19] Scott PB. Wind- PV Hydrogen by Electrolysis. ISE Research – ThunderVolt Inc; April 10, 2003. http://www.espcinc.com/library/15_MW_CAES_Project_with_Above_ground_Storage.pdf
[20] http://thefraserdomain.typepad.com/energy/2006/04/pem_fuel_cells.html.
[21] *www.energystorageandpower.com/pdf/15_MW_CAES_Project_with_Above_ground_Storage.pdf.
[22] Huntorf CAES: more than 20 years of successful Operation, www.unisaarland.de/fak7/fze/AKE_Archiv/.../AKE2003H03c_Crotogino_ea_HuntorfCAES_CompressedAirEnergyStorage.pdf.
[23] *Schainker, 1997 (reproduced in PCAST, 1999). Toward-optimization-of-wind-compressed-air-energy storage.
[24] *Beacon power operating manual of flywheel storage, *www.upei.ca/ physics/p261/projects/flywheel1/flywheel1.htm.

[25] *NEWSWISE, Flywheel batteries come around again, www.newswise.com/articles/ 2002/3/FLYBATT.IEE.html.

[26] *http://en.wikipedia.org/wiki/Protonic_ceramic_fuel_cell.

[27] *US Department of Energy, www.eia.doe.gov/pub/international/iealf/table26.xls.

[28] http://mcg.8m.net/tute/miniHydro/miniHydro.htm.

[29] National Fuel Cell Research Center, www.nfcrc.uci.edu/2/FUEL_CELL_INFORMATION/ FCexplained/FC_Types.aspx; 2012 [accessed 14.10.12].

[30] *http://en.wikipedia.org/wiki/Water_turbine; [accessed 14.10.12].

[31] Rise Research Institute for Sustainable Energy, * www.rise.org.au/info/Tech/hydro/large. html.

[32] How stuff works?, *www.howstuffworks.com/hydropower-plant.htm/printable.

[33] *The Honda FCX ultra capacitor, http://world.honda.com/FuelCell/FCX/ultracapacitor/ charging/.

[34] *www.fuelcellstore.com/en/pc/viewPrd.asp?idproduct=469.

[35] *A comparison of energy storage technologies, www.electricitystorage.org/tech/ technologies_comparisons.htm.

[36] www.odec.ca/projects/2010/hegdxa2/SuperconductingMagneticEnergyStorage.html2012 [accessed 14.10.12].

[37] *http://superconductors.org/; [accessed 14.10.12].

[38] *http://en.wikipedia.org/wiki/Superconductivity (10.14.2012).

[39] *www.electricitystorage.org/technology/storage_technologies/technology_comparison.

[40] *www.accel.de/pages/2_mj_superconducting_magnetic_energy_storage_smes.html.

[41] *www.lanl.gov/superconductivity/; [accessed 14.10.12].

[42] *www.amsc.com/products/library/**DSMESd**s.pdf; [accessed 14.10.12].

[43] *www.ideafinder.com/history/inventions/battery.htm; [accessed 14.10.12].

[44] www.daviddarling.info/encyclopedia/M/AE_microhydropower.html.

[45] http://science.nasa.gov/science-news/science-at-nasa/2003/05feb_superconductor/.

[46] www.csaasia.org/english/news/newsletters/certification_news/winter2002/Default.asp? load=6621; [accessed 14.10.12].

[47] www.energyeducation.tx.gov/technology/section_3/topics/applications_for_fuel_cells/ index.html; [accessed 14.10.12].

[48] http://inhabitat.com/green-technology/page/24/; [accessed 14.10.12].

[49] http://machinedesign.com/article/fuel-cells-long-and-winding-road-0927 (10.14.2012).

[50] www.mathworks.com/company/user_stories/Plug-Power-Accelerates-Fuel-Cell-Control-Development.html; [accessed 14.10.12].

[51] www.sciencedirect.com/science/article/pii/S1464285907703991; [accessed 14.10.12].

[52] www.nasa.gov/centers/dryden/news/ResearchUpdate/Helios/index.html;   [accessed 14.10.12].

[53] www.hec.calpoly.edu/projects/cart/; [accessed 14.10.12].

[54] http://integrating-renewables.org/integrating-renewables-technology-solutions/; [accessed 14.10.12].

[55] www.accessscience.com/loadBinary.aspx?filename=233100FG0040.gif; [accessed 14.10.12].

[56] Sci Tech Rev April 1996–59. www.evergreencpusa.com/products/Rechargeable_NiCd. html; [accessed 30.10.12].

[57] http://ucs.berkeley.edu/energy/category/energystor/; [accessed 30.11.2012].

Emerging Renewable Energy Sources

5.1 OCEAN THERMAL ENERGY CONVERSION

The United States is the world's largest user of energy. Most energy production is based on fossil fuels: oil, gas, and coal—resources that are eventually going to run out. Since production of energy goes up year by year, the rate at which these fuels are burned is also increasing. It therefore is necessary that we invest time and money into researching and developing renewable energy sources. Such methods need to be economically viable and be able to produce energy on a significant scale. The two largest sources for renewable energy are solar and wind.

Wind power plants harness the wind's energy to create rotational energy that is then turned into electricity using a generator. The most widespread form of solar energy conversion is with the use of photovoltaic (PV), or solar panels. These silicon-based panels directly convert solar energy into electricity. This form of energy requires covering large areas with costly solar panels. An alternate approach involves using existing surface area that is absorbing solar energy anyway. This method involves using the energy that the ocean absorbs and turning it into electricity. By using the sun-heated surface water, and the cold water from the bottom of the ocean, methods have been created to turn this temperature difference into electricity. This method is referred to as ocean thermal energy conversion (OTEC).

5.1.1 Introduction

Ocean thermal energy conversion is an idea that has been around for a long time. Several plants have been constructed and operated during the last century. Some research has been done but much more is needed. In its current state, OTEC does not appear to be the most economical choice for producing alternative energy. As more research is completed, OTEC facilities may be built that have enough longevity to make it a viable energy source. OTEC has other features that make it attractive beyond just as an energy source. A byproduct of OTEC is desalinated water for drinking, and the cold water needed to produce the energy can be used for a variety of maricultures such as clams, lobster, and spirulina. Figure 5.1 shows an artist's picture of a typical OTEC plant for generating electricity and other applications mentioned here.

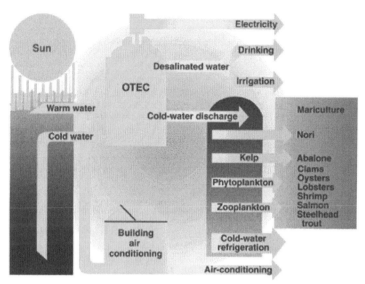

FIGURE 5.1 An artist's view of an OTEC plant.

An ocean thermal energy conversion system needs to be used in an area where there is easy access to waters that have a 20 °C difference. Generally, conversion is done by pumping in warm surface water, and the cold water is pumped up from around 1000 m below the surface. This wide temperature difference largely exists in the tropics.

5.1.2 OTEC schemes

In general, an OTEC plant consists of: a generator to generate electricity, a turbine to drive the generator, an evaporator to generate vapor with high pressure to drive the turbine; the evaporator uses the warm surface water, a condenser to condense the vapor using cold water from the deep, and pumps to circulate the water. There are three basic schemes for converting the temperature difference into energy: open cycle, closed cycle, and hybrid.

Open cycle

The open-cycle scheme (Figure 5.2) follows these steps:

1. Warm seawater is flash evaporated using pressure differences.
2. The water vapor rotates the turbine to generate power.
3. Cold seawater is used to condense the vapor back to water.
4. Water is then recompressed to its original pressure.

The turbine, compressor, and evaporator all operate at about 1 to 3% of atmospheric pressure. This means that care must be taken to prevent leaks. Ocean water contains gases, such as oxygen, nitrogen, and carbon dioxide, which are uncompressible, so

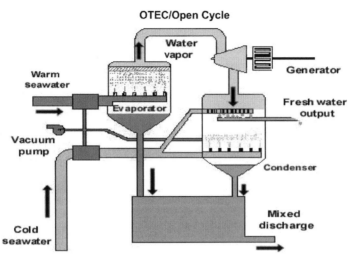

FIGURE 5.2 Graphic representation of an open-cycle OTEC system.

steps must be taken to remove them. A side benefit of the water after it has been compressed is that it has no salt. This is especially a benefit because most potential OTEC locations are on islands that have little natural fresh water.

Closed cycle

A closed-cycle OTEC is when warm seawater is pumped through a heat exchanger that vaporizes a liquid with a low-boiling point (e.g., ammonia). The expanding ammonia vapor is used to rotate a turbine that makes electricity. The ammonia vapor is then condensed by the cold water and sent back to the first heat exchanger. This process is shown in Figure 5.3. Several closed-cycle OTEC plants have been built.

Hybrid OTEC plants are a third methodology. Hybrid is just a facility that uses both open- and closed-cycle converters.

5.1.3 Power and efficiency calculation

Carnot efficiency

In a Carnot cycle, there is a heat source (e.g., flame, sunlight) and a cooling source (e.g., water, cold air). The definition of Carnot efficiency is η:

$$\eta = 1 - \text{Temp.Cold}/\text{Temp.Hot} \tag{5.1}$$

where temperature is in Kelvin. Meanwhile, real efficiency is given as:

$$\eta = 1 - \sqrt{(\text{Temp.Cold}/\text{Temp.Hot})} \tag{5.2}$$

In an idealized system with perfect heat exchangers, the volume flow, Q, of warm water passes into the system at temperature, T_h, and leaves at T_c (the cold water

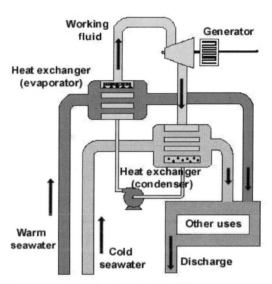

FIGURE 5.3 Graphic representation of a closed-cycle OTEC.

temperature of lower depths). The output power given up from the warm water in such an ideal system is:

$$P_o = \eta_{tg} C_p Q \Delta T \qquad (5.3)$$

where,

    P_o = power given by the warm water
    Q = water flow, kg/s
    $\Delta T = T_h - T_c$ in Kelvin, K
    C_p = specific heat of water
    C_p = 4.187 kJ/kgK
    η_{tg} = turbine generator efficiency

EXAMPLE 5.1

a) What is the theoretical and real efficiency of an OTEC if the surface ocean temperature is 80 °F (299.6 °K) and the deep water temperature is 50 °F (283 °K)?
b) Calculate the power output of the system if the turbine generator efficiency is 85% and the flow rate of both the warm and cold water is Q = 6100 kg/s.

Solution

a) Using the Carnot efficiency equation: 1-cold/hot

$$\eta = 1 - 283/299.6 = 6\% \text{ theoretical effiency}$$

EXAMPLE 5.1—cont'd

Using the Carnot efficiency equation for irreversible cycle:

$$1 - \sqrt{(\text{cold/hot})}$$
$$\eta = 1 - \sqrt{(283/299.6)} = 2.8\% \text{ real efficiency}$$

Which is similar to what was achieved in an OTEC plant at Keahole Point in Hawaii (1 – 3%)?

b) The thermal energy in water is (flow rate) × (temp) × (C_p). At 299.6 K, the warm water equation is:

$$(6100 \text{ kg/s})(299.6 \text{ K})(4.187 \text{ kJ/kgK}) = 7,665,200 \text{ kJ/s}$$

At 283 K, the cold water carries

$$(6100 \text{ kg/s})(283 \text{ K})(4.187 \text{ kJ/kgK}) = 7,228,000 \text{ kJ/s}$$

The maximum energy extracted is the difference between the two:

$$7,665,200 - 7,228,000 \text{ kJ/s} = 423,975.62 \text{ kJ/s}$$

with a turbine generator efficiency of 85%. The real power output is:

$$P_o = 423,975.62 \text{ kJ/s} \times 0.85 = 360379.27 \text{ kJ/s} = 360,379.27 \text{ kW}$$

5.1.4 Land-based and floating OTEC plants

The first literary reference to an OTEC was in Jules Verne's *20,000 Leagues under the Sea* published in 1870. The OTEC can be constructed floating offshore (Figure 5.4) or as a land-based facility. The federal government initiated a program to develop OTECs and started with the construction of miniature one in 1979 (Figure 5.5). The mini-OTEC was the offshore-type and produced 50 K. This plant was only operated for a few months.

In 1993, a 210-kW closed-cycle OTEC plant was built in Hawaii (Figure 5.6). This land-based plant was successfully operated for six years (1993–1998). While in operation, 10% of the steam this plant produced was diverted to a desalination device for the production of drinking water for the island. Another floating OTEC plant that was designed by the federal government was OTEC-1 (Figure 5.7). OTEC-1 is a major part of the Department of Energy's (DOE) developmental program for ocean thermal energy conversion. It is a floating test facility designed primarily for conducting sea tests of the cleanability and performance of heat exchangers, a major component of OTEC systems.

Each of these options has its own advantages and disadvantages. Floating facilities are closer to the deep waters that have the temperature necessary, and therefore do not need to have as long a pipe as a land-based counterpart. However, floating plants face the difficult challenge of transporting the energy back to shore.

FIGURE 5.4 A Photo of the 1979 mini-OTEC plant.

FIGURE 5.5 Land-based 210 kW OTEC plant.

5.1.5 **Advantages and disadvantages of OTEC**

Besides generating electricity, there are other uses for OTEC plants; one is because they generate desalinated water for human consumption. The cold water that is pumped up from the bottom of the ocean is rich in nutrients not normally found on the surface. So, instead of pumping the water back into the ocean, it can be used to support all kinds of mariculture.

OTEC 1

FIGURE 5.6 An artist's representation of OTEC-1.

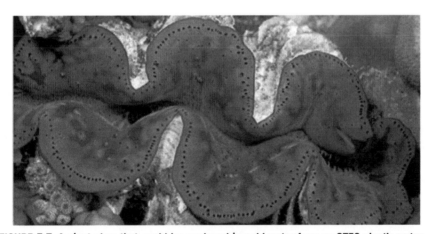

FIGURE 5.7 A giant clam that could be produced in cold water from an OTEC plant's water.

Advantages

The cold water can be pumped into large vats with giant clams, lobsters, flounder, spirulina, and all kinds of sea vegetation; for example, see Figure 5.8. Since the mariculture is close by, it will be much less expensive to produce and harvest since the cold water is being used anyway. The cold water could also be used for refrigeration, or it could have its minerals extracted for traditional farms.

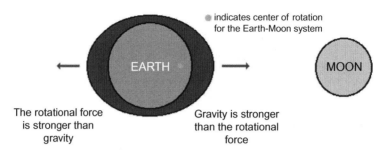

FIGURE 5.8 Effects of the rotational and gravitational forces of the Moon on the Earth.

Ocean thermal energy conversion uses clean, renewable resources. All the things it needs are readily available. There is enough energy stored in the ocean to produce all the energy that we currently need. Well-designed OTEC plants will not produce carbon dioxide or other greenhouse gases that are harmful to the environment. Use of OTEC will decrease our dependence on fossil fuels.

Disadvantages

OTEC is expensive; it costs much more than fossil fuels and more than other sources of renewable energy. The truth of the matter is: given the low cost of fossil fuels, and the availability of inexpensive renewable resources, the high cost of OTEC makes it not viable economically.

The less costly plants are the ones that operate on a ship. This means that a method needs to be conceived to transport that energy to the land. Some possibilities include hydrogen production or submarine cables; the economical option right now is cables. If a plant could operate for more than 30 years without a major overhaul, then the cost of OTEC could go down substantially. However, there is little data showing how long we can expect an OTEC plant to operate.

Land-based OTEC plants need to have access to a section of ocean that has a greater than 20-degree temperature difference between the surface and the depths. There are relatively few places where this occurs (relative to tidal, solar, or wind power). Alternatively, we could build floating ocean-based plants; however, this scenario involves the high cost of transporting the energy to shore.

Even though smaller plants have been built and operated, it is necessary to build and operate a commercial-sized plant for a long period of time. Without doing this, we have no way of properly assessing the viability of OTEC. Construction of large pipes could adversely affect local marine ecosystems and reefs. Additionally, there may be negative effects from the pumping of large volumes of cold water into an area that is generally warm.

Additional component research is needed because some of the parts that an OTEC system relies on are not fully developed, including large-diameter deep sea pipes, low-pressure turbines, and condensers. The efficiency of OTEC plants is very low at about 2%.

5.1.6 **The future of OTEC**

Several designs have been produced, but because of limited funding, it is doubtful that any of the designs will ever be produced. One proposed plant floats and could produce around 5 MW of electricity. The plant would draw cold water through a 1000-m pipe and use ammonia in a closed-cycle plant to generate the electricity (see Section 5.1.4).

The future of OTEC is unstable at best. It is difficult to spend lots of money on this technology when there are many others that are much closer to being an economic and a technical reality. Wind power can be used in any area that an OTEC can be. Ocean thermal energy conversion needs to be close to the shore and coastlines are always windy due to the temperature difference between land and water. Wind power is much more developed and can be purchased and operated for much less than an OTEC plant could be. Since OTEC works mostly in the tropics, these areas are also good candidates for PV solar power.

It is this author's belief, based on reading the literature, that OTEC may eventually be a viable alternative energy source, but right now there are many superior choices.

5.2 **TIDAL ENERGY**

Tidal energy is the utilization of the gravitational forces of the Sun and Moon—tides are formed by the pull on the oceans of the rotating Earth. Tides can be found with varying degrees of strength on any coastline.

5.2.1 **Background on tides**

Tides are changes in ocean levels generated by the gravitational interactions between the Sun, Moon, and Earth. Each body is pulling on the other; consequently, the oceans on the Earth move as the various interactions become stronger or weaker. Newton's law of gravitation can be used to directly determine the gravitational interactions. In the attraction between the Earth and the Moon, the gravitational force is balanced by the centrifugal force of the two bodies rotating about their common center of gravity.

The balance is created at the center of mass for both bodies; however, at the surface of the Earth, there is an imbalance. This results in a tide-producing force on the side of the Earth facing the Moon. There is also a smaller tide-producing force on the Earth's opposite side. Each of these forces creates a bulge along the axis between the Earth and the Moon similar to that shown in Figure 5.9. These forces create a cycle of approximately half a day, known as the "lunar tide."

In addition to the Moon's effect on the Earth's oceans, the Sun produces a similar effect, known as the "solar tide." Although the Sun is a much larger body, the distance between the Sun and the Earth counteracts the effect, and the resultant bulge in the ocean due to the Sun is less than that of the Moon. The effect of the Moon is approximately 2.2 times greater than the effect of the Sun.

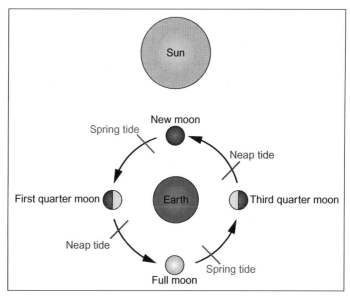

FIGURE 5.9 The phases of the tides.

Due to the different orbits of rotation for the Sun and the Moon, the tidal bulges for each occur at different frequencies. Like all other waveforms, these bulges can act either constructively or destructively. When the tides are in-phase, the "spring" (maximum) tides occur; when the tides are out-of-phase, the "neap" (minimum) tides occur. Figure 5.10 shows the phases of the tides.

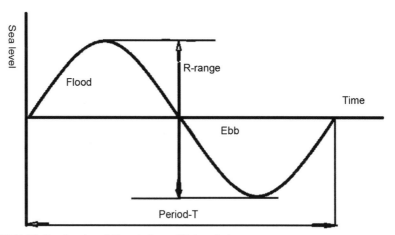

FIGURE 5.10 Sea-level variation during a tidal period.

5.2.2 Tidal periodicity, amplitude, and power output calculation

Each of the interactions between the Sun, Moon, and Earth cause the surface of the oceans to rise and fall in a periodic fashion according to the force given by Newton's law of universal gravitation. The upward-acting lunar gravitational force results in cyclical variations in sea level, sinusoidal with 12.4-hour cycle (Figure 5.11). These interactions are cyclic with the nature of the rotation of each body. In summary, several different cycles can be seen in the tides including

- A 12.4-hour cycle due to the rotation of the Earth within the gravitational field of the Moon.
- A 14-day cycle due to the gravitational field of the Moon combining with that of the Sun to give alternating spring tides and neap tides.
- A half-year cycle due to the inclination of the Moon's orbit to that of the Earth, giving rise to maximal tides in March and September.

After discussing the periodic nature of the tides, one may wonder at the magnitude to which they actually change. In the open ocean, the difference between tide levels is less than 1 m; however, tidal amplitudes can be increased significantly near the shoreline due to several different effects including shelving, funneling, reflection, and resonance. Shelving causes the tidal amplitude to increase as the depth of the ocean decreases. Based on the coastline, funneling may amplify the tide if the shape of the coastline becomes increasingly narrow. Reflection of waves from the coastline, adding in a constructive manner, can increase tidal amplitude. Resonance occurs when the driving tide at the mouth of an estuary is a multiple of the natural frequency of the tidal propagation into the estuary.

The following equation governs the maximum output power from a tidal system:

$$P = \frac{\rho g R^2 S}{2T} \tag{5.4}$$

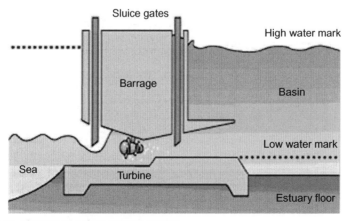

FIGURE 5.11 Diagram of a tidal barrage system.

where,

P = power generated (in watts)
ρ = density of the water (seawater is 1025 kg/m^3)
g = gravitational acceleration
R^2 = tidal range
S = surface area of the tidal basin
T = period (12.4 hr)

The formula in Eq. (5.4) shows that the power out of a tidal system varies directly with the square of the tidal range and with the surface area of the basin. Therefore tidal range is a very important factor in the choice of location for a tidal power plant.

EXAMPLE 5.2

Calculate the power output generated by the Bay of Menzen, part of the White Sea, in Russia. Use Table 5.1 later in the chapter and Eq. (5.4) to solve this problem. Assume the surface area of the tidal basin is $S = 5 \times 10^6$ m^2 and the tidal range is $R = 10$ m.

$$P = [(1025.18 \text{ kg/m}^3)(9.81 \text{ m/s}^2)(10 \text{ m})^2(5 \times 10^6 \text{ m}^2)/(2)(12.4)(3600)\text{s}]$$
$$= (56 \text{ MW})(60\%) = 34 \text{ MW}$$

5.2.3 Tidal energy conversion systems

The generation of electricity from the tides is very similar to hydroelectric generation, except that water is able to flow in both directions, and this must be taken into consideration in the development of the generators. Basically, there are two methods for generating electricity using the tide's power:

- Tidal barrage, which consists of a storage pond that is filled with the flood tide and emptied on the ebb tide. An incoming tide is called a "flood tide" and an outgoing tide is called an "ebb tide." It also consists of a generator–turbine unit and gates and ducts to funnel water through.
- Tidal current method is where tidal turbines use currents that are moving with velocities between 2 and 3 m/s (4–6 knots) to generate between 4 and 13 kW/m^2. Fast-moving current (>3 m/s) can cause undue stress on the blades in a similar way to that of very strong gale force winds, which can damage traditional wind turbine generators, while lower velocities are not economical.

Tidal barrage power plants

The tidal barrage is the most commonly used method to harness tidal power. It consists of sluice gates and turbine generators built into a barrage (dam) across the mouth of a river, estuary, or coastal bay (Figure 5.12). Several different turbine configurations are

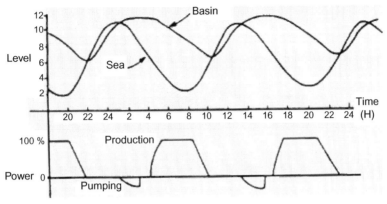

FIGURE 5.12 The ebb-generation mode.

possible. For example, the La Rance Barrage tidal plant near Saint-Malo on the Brittany coast in France uses a bulb turbine. In systems with a bulb turbine, water flows around the turbine, making access for maintenance difficult because the water must be prevented from flowing past the turbine. Rim turbines, such as the Straflo used at the Annapolis Tidal Generating Station in Nova Scotia, reduce these problems because the generator is mounted in the barrage at right angles to the turbine's blades. Unfortunately, it is difficult to regulate the performance of these turbines, and they are unsuitable for use in pumping.

Tubular turbines have been proposed for use in the Severn Barrage tidal project in the United Kingdom. In this configuration, the blades are connected to a long shaft and oriented at an angle so that the generator is sitting on top of the barrage.Tubular ducts within the barrage contain the turbines, which drive electrical generators. The sluice gates and the turbine ducts are opened to let the flood tide fill or empty the basin behind the barrage. There are three modes of operation for generating electricity, as follows:

Ebb generation: Sluice gates on the barrage allow the tidal basin to fill on the incoming high tide and exit through the turbine system on the outgoing tide. At high tide, when the basin is filled, the sluice gates are closed. They stay closed for a period of time, allowing a difference of water levels between the basin-side and sea-side of the barrage as the ebb tide goes out. The sluices are reopened to let water from the basin spill through the turbine ducts and over the turbine, generating electricity. This operation is very similar to a low-head hydroelectric dam. The example tidal barrage described is operating in a mode called ebb generation; it is shown in Figure 5.13.

Flood generation: This is electricity generation on the incoming tide rather than the outgoing (Figure 5.14). Experience in the United Kingdom's Tidal Energy

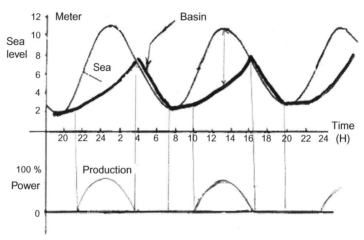

FIGURE 5.13 The flood-generation mode.

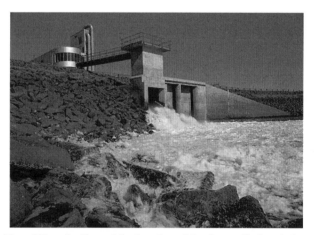

FIGURE 5.14 Map of Nova Scotia and the location of the Annapolis Royal plant.

Programme has shown that ebb generation maximizes the amount of energy that can be produced in a barrage-type system.

Two-way generation: This mode uses both tides, ebb and flood, to produce electricity.

Sites of tidal barrages and feasibility

Tidal power requires large tidal differences. There are a number of sites around the world that have the right combination of coastal geometry and a significantly large tidal range to make them potential sites for tidal power plants like the next subsection describes.

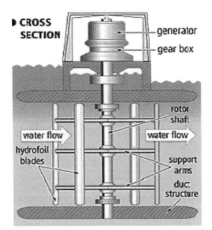

FIGURE 5.15 The Darrieus turbine built by Blue Energy Ltd.

The annapolis tidal generating station in Canada

Construction for the generating station began in 1980 and was completed four years later. It is the only modern tidal generating plant in North America. The plant is located on a small island at the mouth of the Annapolis River off of the Bay of Fundy in Nova Scotia (Figure 5.15). The Bay of Fundy is known for having some of the highest tides in the world, with an average tidal range in the Minas Basin of approximately 12 m. The tidal range at Annapolis Royal averages 7 m.

The plant houses two sluices and one Straflo turbine. On the flood tide, the sluice gates are opened to allow incoming seawater to fill the head pond. When it is filled to its maximum level (below high tide in Annapolis Royal), the sluice gates are closed, trapping the water in the head pond upstream of the turbine. The plant then waits until the sea level has receded enough to begin power generation. This occurs when the difference in the level between the head pond and sea level is 1.6 m or more.

The economic viability of a tidal power plant using a barrage design includes consideration of startup cost, design, and power production. The initial investment necessary to build a tidal barrage is a lot. However, once the plant is built, minimal maintenance is required. There are only four relatively large tidal plants in existence; they are described in Table 5.1, not many examples to build on. Each design is also site-specific and must take into account the need for potential ship locks if navigation through the barrage is required; this would increase the initial investment.

Advantages and disadvantages of tidal barrages

The construction of the dam and the installation of the turbine generator are very costly. Economic considerations must also include the actual power production cycle of a tidal power plant. The production cycle shows that in a 24-hour period, an ebb-tide system will only generate electricity for approximately 12 hours, which

Table 5.1 Existing Tidal Schemes Throughout the World

	Completed	Tidal Range	Basin Area	Capacity
La Rance, France	1966	8.0 m	17 sq. km	240 MW
Kislaya Guba, Russia	1968	2.4 m	2 sq. km	0.4 MW
Jiangxia, China	1980	7.1 m	2 sq. km	3.2 MW
Annapolis Royal, Canada	1984	6.4 m	6 sq. km	17.8 MW
Eight more in a China	1961–89	1.2–3.5 m	—	—

means significant periods of time with no power output, and the 12 hours of generation shifts every day by one hour. So even though the cycle is very reliable, it does not correlate with the typical energy demands during a day. This energy mismatch means that production by tides cannot be a sole source of power for a region unless energy storage is employed. This too will drive the price of the tidal plant higher.

Environmental viability is linked to the impact of the tidal power system on the surrounding ecosystem. By introducing an unnatural barrier, there will be an increase in sedimentation behind the barrage, as well as an increase in the salinity of the water. Fish population would be affected since estuaries are key to many migratory fish. Wading birds would also be affected since the percentage of mudflats exposed at low tide would drop. The cost associated with these environmental impacts is difficult to determine, however, studies are underway.

Tidal current power plants

Tidal turbines proposed shortly after the oil crisis of the 1970s have only become a reality in 1994, when a 15-kW proof-of-concept turbine was operated on Loch Linnhe in Scotland. Resembling a wind turbine, tidal turbines offer significant advantages over barrage and fence tidal systems, including reduced environmental effects. Tidal turbines use currents that are moving with velocities of between 2 and 3 m/s (4–6 knots) to generate between 4 and 13 kW/m^2. Fast-moving current (>3 m/s) can cause undue stress on the blades. Tidal current models do not require the construction of an expensive dam.

Tidal current technologies

Several methods built by the following companies are used to generate electricity:

- Blue Energy Ltd—Darrieus turbine
- Marine Current Turbines Ltd—Vertical turbine
- UEK Corporation—Underwater electric kite
- ANDRITZ HYDRO Hammerfest Strøm—Vertical turbine
- The Engineering Business Ltd—Stingray generator

FIGURE 5.16 Vertical turbine built by Marine Current Turbines Ltd.

Blue energy's Darrieus turbine. Blue Energy's larger ocean class Darrieus Turbine is 10.5 m in diameter, is able to generate a peak power of 14 MW, and can operate at depths up to 70 m (Figure 5.16). Multiple Davis Hydro Turbines can be mounted in a modular duct structure many kilometers long to form a tidal fence or tidal bridge across a river, tidal estuary, or ocean channel. The superstructure of a typical tidal fence can be used as the base for a bridge, or as a platform for other commercial developments such as offshore wind turbines.

It can also carry utility piping and wiring such as water mains, gas lines, phone lines, fiber optics, and power lines. The tops of the machinery rooms are continuous and support crane rails and can have a double-lane roadbed for vehicular traffic, either along the top of the tidal fence or enclosed within the structure. Modules are 2 kW, 5 kW, and 250 kW for river applications and can be fixed to the ocean bed or held in place by cables.

Marine current's vertical turbine. Marine Currents's vertical turbines are similar to a submerged windmill. The subocean turbines consist of twin-axial flow rotors 15 to 20 m in diameter, each driving a generator through a gearbox similar to a wind. The twin power units of each system are mounted like extensions on either side of a tubular steel piling 3 m in diameter and are set into a hole drilled into the seabed (Figure 5.17). They are rated from 1 to 2 MW. The turbines can be grouped in arrays similar to wind turbines in a windfarm and can be installed in water depths of 30 to 100 m.

UEK's underwater electric kite. The structure of the UEK is a self-contained, moderately buoyant turbine/generator that is suspended like a kite within the tidal stream. Underwater electric kite turbines can be anchored to the seabed or attached to existing structures (e.g., a bridge) or put on a floating platform (Figure 5.18). If the depth of the water is substantial and the current flow is not confined, the kite can be tethered to an electro-anchorage cable that allows the kite to position itself in the fastest part of an ocean or river current. The 3.0-m twin turbine delivers 90 kW in 5-knot currents. UEK states that they will have a 6.7-m twin turbine system available in the near future and continue to work to design a 1-MW twin kite turbine.

FIGURE 5.17 Underwater electric kite built by UEK Corporation.

FIGURE 5.18 The vertical turbine built by Dhammerfest Strøm.

Dhammerfest Strøm 's vertical turbine. This underwater windmill has three 10-m blades that rotate at 5.15 rpm and can produce 300 kW of electricity (Figure 5.19).

Engineering business's Stingray. It uses an oscillating wing-like hydroplane (Figure 5.20). As tidal currents pass over the hydroplane, lift and drag forces cause it to rise. The 250-kW hydroplane stands 23.6 m tall, has an arm length of 11 m, and a wing span of 15.6 m. A 500-kW version of the Stingray is in the works.

5.3 WAVE POWER GENERATION SYSTEMS

Some of the solar energy received by the Earth is converted to wind energy, and some is converted to wave energy using a variety of wave power generation systems (WPGS). Waves are a free and sustainable energy resource created as wind blows over the ocean's surface. The greater the distances involved, the higher and longer the waves will be.

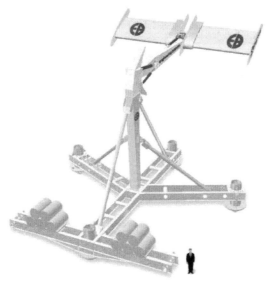

FIGURE 5.19 Stingray built by The Engineering Business Ltd.

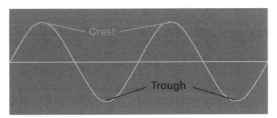

FIGURE 5.20 A wave's profile.

5.3.1 **Wave power principles**

Waves are sinusoidal in nature (Figure 5.21). Energy is stored in this way until it reaches the shallows and beaches of the coasts where it is released, sometimes with destructive affects. The crest is the highest part of the wave and the trough is the lowest (Figure 5.22). The distance between the crest and the trough is the wave height. The distance from the crest to crest is the wave length, λ. The period of a wave is the time it takes for each crest to pass a certain point.

The power in a wave is equal to the product of the potential energy per unit area times the phase velocity of the wave (c). Linear wave theory assumes that the motion of the water past a certain point is sinusoidal. The period, T, for one wave to pass this point can be expressed as:

$$T = \sqrt{\frac{2\pi.\lambda}{g}} \qquad (5.5)$$

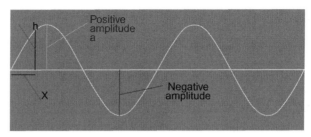

FIGURE 5.21 Graphic of amplitude as a function of distance.

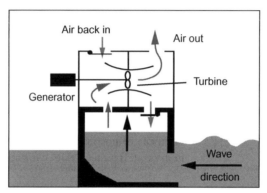

FIGURE 5.22 The OWC system.

where,

g = gravity
ρ = water density
λ = the wave length

The phase velocity, or celerity, is:

$$c = \frac{gT}{2\pi} \qquad (5.6)$$

The potential energy per unit area is:

$$E = 0.5\,\rho g a^2 \qquad (5.7)$$

The power contained in the wave can be expressed in terms of the length of the wave (kW/m). This is given by the following equation:

$$p = \frac{\rho \cdot g^2 \cdot a^2 \cdot T}{8\pi} \qquad (5.8)$$

where a is the peak amplitude of the wave.

Wave size is a factor of wind speed and fetch, the distance over the ocean's surface that the wind travels. The World Energy Council estimates that an amount of energy could be obtained from the oceans that is equivalent to twice the electricity produced in the world. The energy's density can average 65 MW per mile of coastline.

EXAMPLE 5.3

A wave in deep water has a time period of 10 sec and a height of 0.91 m.

(a) Calculate the wavelength.
(b) Calculate the phase velocity.
(c) Calculate the total potential energy.
(d) Calculate the power.

For deep water,

Wavelength:

$$\lambda = \frac{gT^2}{2\pi} \quad \lambda = \frac{9.8 \times 10^2}{2\pi} = 155.972 \text{ m}$$

Phase velocity:

$$c = \frac{gT}{2\pi} \quad c = \frac{9.8 \times 10}{2\pi} = 15.597 \text{ m/s}$$

where c is $H = 2a$.

The total energy in the wave is:

$$E = \frac{\rho g \lambda H^2}{8}$$

$$E = \frac{1030^*9.8^*155.972^*(0.914)^2}{8} = 164.403 \text{ kJ}$$

Wave power:

$$P = \frac{\rho g H^2 c_g}{8}$$

For deep water,

$$c_g = \frac{c}{2} \quad c_g = \frac{15.597}{2} = 7.799 \text{ m/s}$$

$$P = \frac{1030^*9.8^*(0.914)^{2^*}7.799}{8} = 8.22 \text{ kW}$$

5.3.2 Types of WPGS

The purpose of a wave energy device is to harvest the energy in the waves and convert it into electricity. There are different types of WPGSs. Some systems extract energy from surface waves. Others extract energy from pressure fluctuations below the water's surface or from the full wave. A wave energy converter can be placed in the ocean in various situations and locations. It may be floating or submerged completely in the sea offshore, or it may be located on the shore or on the seabed in relatively shallow water. The following are most of the devices used to generate electrical power from ocean waves:

- Oscillating Water Column (OWC)
- Surface followers using floats or pitching devices
- Focusing devices
- Submarine turbines
- Mighty Whale floating OWC
- Solar water pump

Oscillating water column

The Oscillating Water Column depends on the moving waves to generate electricity in a two-step process; that energy can be used to power a turbine. As a wave enters the column, it forces the air up the closed column past a turbine, which increases the pressure within the column. As the wave retreats, the air is drawn back past the turbine due to the reduced air pressure on the turbine's ocean side.

In this simple example, the wave raises into a chamber. The rising water forces the air out of the chamber. The moving air spins a turbine, which can turn a generator. When the wave goes down, air flows through the turbine and back into the chamber through doors that are normally closed (Figure 5.23). Figure 5.24 shows a schematic diagram of an OWC. The energy of this air stream is extracted to generate electricity through the use of a pneumatic turbine. The OWC's turbine is able to turn in the same direction without regard to which way the air is flowing across its blades. Therefore the turbine continues to turn on the rise and fall of the water inside the chamber, increasing efficiency.

Limitations of the OWC

In these devices, the water oscillates up and down in a fixed pipe. The pressure variation of the water surface underneath the pipe forces the air above the surface in the pipe to oscillate. This oscillating airflow can then be passed through a turbine. The horizontal length of the pipe has to be smaller than one-quarter of the wavelength or the efficiency drops markedly, reaching zero when the length is one wavelength. Other limitations on the amount of energy that can be extracted are that these devices rely on the compression/expansion of a body of air, causing heat losses.

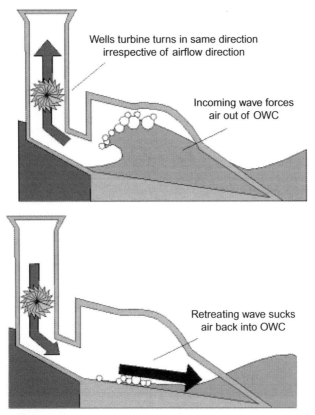

Wells turbine turns in same direction
irrespective of airflow direction

Incoming wave forces
air out of OWC

Retreating wave sucks
air back into OWC

FIGURE 5.23 Schematic diagram of an Oscillating Water Column.

Surface followers

Surface followers, or floating devices, use mechanical linkage between two floating objects and a fixed object to produce useful mechanical power. This power can either be connected directly to a generator or transferred to a working fluid, water or air, that drives a turbine (Figure 5.25). The two main surface following devices are the Cockerel raft and the Salter duck.

The Cockerel raft is comprised of a series of rafts connected together by hinges. Hydraulic pistons can also connect the raft sections that generate the energy. "The Salter Duck, Clam, Archimedes wave swing and other floating wave energy devices generate electricity through the harmonic motion of the floating part of the device, as opposed to fixed systems, which use a fixed turbine powered by the motion of the wave. In these systems, the devices rise and fall according to the motion of the wave and electricity is generated through their motion.

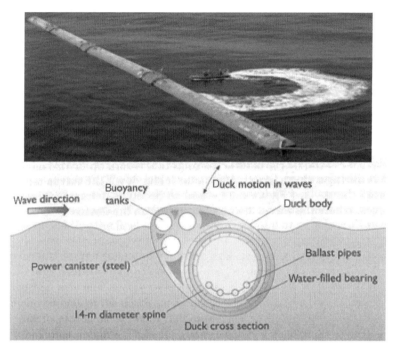

FIGURE 5.24 A surface follower WPGS.

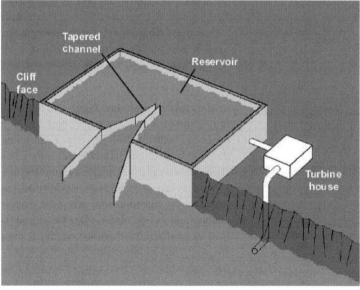

FIGURE 5.25 A TAPCHAN WPGS.

The Salter Duck is able to produce energy extremely efficiently; however, its development was stalled during the 1980s due to a miscalculation in the cost of energy production by a factor of 10. Only in recent years has it been that the technology was reassessed and the error identified."

Limitations of floating devices. Many "up and down" devices use a float; the movement up and down with the wave is used to create a pushing/pulling force in relation to a fixed point—either an anchoring block or a damping plate. The possible amount of extracted energy in these devices can never be greater than the potential energy component of the wave. Floats also have size limitations: If the horizontal length of the float perpendicular to the wave front is larger than one-quarter of the wavelength, the efficiency drops because the float starts to roll about its center of gravity. A float larger than the wavelength will ride on several wave crests without "dropping down" in the wave trough. Also, since wave amplitude decreases the deeper it is, the maximal potential energy of a wave exists only at the wave's surface. As the float gets larger vertically, its center of gravity will be deeper and the potential energy available will be less.

Focusing devices (TAPCHAN)

The TAPCHAN, or tapered channel system, consists of a channel that feeds into a reservoir, which is constructed on a cliff as shown in Figure 5.26. The narrowing of the channel causes the waves to increase their amplitude (wave height) as they move toward the cliff face; eventually they spill over the walls of the channel and into the reservoir that is positioned several meters above mean sea level. The kinetic energy

FIGURE 5.26 A marine turbine WPGS.

of the moving wave is converted into potential energy as the water is stored in the reservoir. The stored water is then fed through a Kaplan turbine. So the concept of TAPCHAN is to collect the water, store it, and run it past a turbine on its way out. There are very few moving parts, and all are contained within the generation system.

Advantages and limitations of a TAPCHAN

TAPCHAN systems have low maintenance costs and a greater reliability. They also overcome the issue of power on demand because the reservoir is able to store the energy until it is required. Unfortunately, TAPCHAN systems are not appropriate for all coastal regions. Locations for TAPCHAN systems must have consistent waves, with good average wave energy, and a tidal range of less than 1 m. Coastal features need to include deep water near to shore and a suitable location for a reservoir.

Marine turbine

This device takes advantage of high tidal current velocities. These devices can be grouped much like turbine windfarms except that a marine turbine for the same rated power is smaller than a wind turbine, and they can be packed closer together. Such devices will be generally 600 to 1000 KW and be less than 50% of the equivalent size wind turbine. A representation of the marine turbine can be seen in the Figure 5.27.

The Mighty Whale

"The world's largest offshore floating wave power device was launched in July 1998 by the Japan Marine Science and Technology Center. The full-scale prototype was to be demonstrated and tested over a two-year period at the mouth of Gokasho Bay

FIGURE 5.27 The Mighty Whale WPGS.

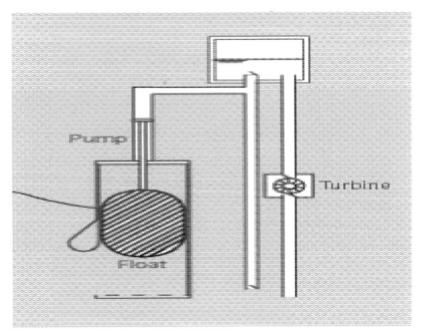

FIGURE 5.28 A wave pump WPGS.

facing the Pacific Ocean." It was expected that this project would show great potential for this type of floating wave power device.

"The Mighty Whale converts wave energy into electricity by using oscillating columns of water to drive air turbines. As shown in Figure 5.28, waves flowing in and out of the air chambers at the mouth of the Mighty Whale make the water level in the chambers rise and fall. The water forces air into and out of the chambers through nozzles on the tops of them. The resulting high-speed airflow rotates air turbines, which drive the generators. The Mighty Whale is 50 m long and 30 m wide and carries three air turbine generator units: one with a rated output of 50 kW + 10 kW and two of 30 kW."

"Because it has absorbed and converted most of the energy in the wave, the Mightly Whale also creates calm sea space behind it, and this feature can be utilized—for example, to make areas suitable for fish farming and water sports. The structure of the Mighty Whale itself can be used as a weather monitoring station, a temporary mooring for small vessels, or a recreational fishing platform. As well as generating energy for use onshore, the Mighty Whale can provide an intermediate energy source for aeration to improve water quality."

A wave pump WPGS

A wave pump is a device that absorbs wave energy by means of a float (Figure 5.29) that oscillates up and down inside a cylindrical structure with a submerged opening through which waves can move. The energy in the waves is converted into mechanical

FIGURE 5.29 The Geysers geothermal power plant in California.

energy by a pump with its piston rigidly connected to the float. Seawater is pumped into energy storage that is in the form of a pressure tank (or possibly an elevated water reservoir) from which a turbine can be run to complete the conversion to electrical energy. Alternatively, a hydraulic fluid, other than seawater, may be used if the system is a closed loop.

Maximum energy capture is obtained in two ways: (1) a control system latches onto the float during certain time intervals of the wave cycle to achieve optimum timing of the float motion and (2) a control means is used to obtain optimum flow of hydraulic fluid into the energy-storage reservoir.

Advantages and disadvantages of wave pumps
The following are some of the advantages and disadvantages of wave pumps:

Advantages
- Renewable source of energy
- Positive environmental effect: replaces fossil fuel energy, cutting down on greenhouse gas emissions and atmospheric pollution
- Free source of energy
- Clean source of energy

Disadvantages
- Waves are irregular in direction, durability, and size, so difficult to harness
- Extreme weather conditions may produce waves of freak, or rouge, intensity
- Submarine cables
- Might have unacceptable visual impacts
- Hurricanes may destroy the energy system and its marine cables

5.4 GEOTHERMAL ENERGY SYSTEMS

The derivation of the word *geothermal* comes from the two Greek words *geo* and *therine*—earth and heat. Geothermal energy is derived from the natural heat of the Earth. Geothermal power plants take advantage of this natural, clean energy source for their geothermal energy systems (GES).

5.4.1 Introduction

Geothermal sources vary by location and depth toward the Earth's core. The temperature at interior is 7000 °C, decreasing to 650 to 1200 °C at depths of 80 to 100 km. This energy is usable from room temperature to in excess of 300 °F. To generate electricity, a geothermal reservoir providing steam and hot water is necessary. The reservoirs can be classified as high temperature (>150 °C) or low temperature (<150 °C). Most commercial installations are located at a site of a high-temperature geothermal reservoir.

Such reservoirs are in regionally localized geologic settings where the Earth's natural heat is located near enough to the surface to allow access to the steam and hot water. Some natural sources of geothermal are located in The Geysers region in Northern California; the Imperial Valley in Southern California; and Yellowstone in Idaho, Montana, and Wyoming. Drilling wells into these reservoirs and piping the steam or hot water to power plants allows the Earth's energy to be converted to electricity.

Geothermal energy has been used by humans for many centuries in applications such as cooking, heating, medicinal bathing, and water heating. The first geothermal power generation plant was built in 1904 in Larderello, Italy. This plant had a capacity of 250 kW and used a direct steam system to generate electricity. The second geothermal power station was constructed in the 1950s at Wairakei, New Zealand. This was followed by The Geysers power plants in California during the 1960s; one is shown in Figure 5.30.

Geothermal is a reliable green energy source, and electricity from geothermal plants is clean. No fossil fuels are burned during the generation process. In the United States, currently more than 2700 MW of electric power is generated by geothermal systems. This is comparable to 60 million barrels of oil per year or enough electricity for 3.5 million homes. This geothermal electricity generation displaces the emission of 22 million tons of CO_2 per year. Geothermal power plants have a system availability of 95% or higher. A comparable coal or nuclear power plant would have a system availability of 60 to 70%. The most important motivation for using geothermal electricity is its cost advantage. Geothermal electricity cost ranges from $0.05 to $0.08/kWh.

Technology improvements supported by the US Department of Energy and the geothermal industry are continuing to reduce this cost to a possible future cost of $0.03/kWh. Plant infrastructure costs are also less than a comparable coal or nuclear plant. An average geothermal power plant requires only 400 m^2 of land to produce a gigawatt of power for more than 30 years. Compare that with the enormous amount

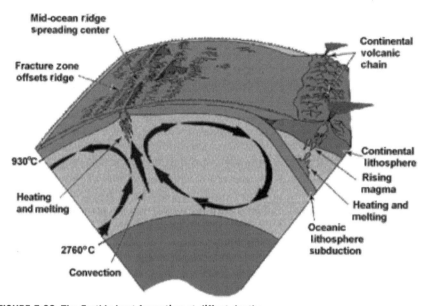

FIGURE 5.30 The Earth's heat formation at diffent depths.

Adapted from: Volcanoes Crucibles of change Princeton University Press, 1997

of land needed for a coal plant and the associated area used in the mines to provide the fuel. The small footprint—1.8 to 7.9 acres/MW geothermal versus 18.9 acres/MW coal, or 5 acres/MW nuclear—make geothermal power plants compatible with scenic areas.

5.4.2 Geothermal resources

Since the beginning of humankind hot springs and thermal pools have been used as indicators of the Earth's shallow heat anomalies. These phenomena as well as geysers, boiling mudpots, and steam vents (fumaroles) are the surface effects of natural hydrothermal systems. Figure 5.31 shows the heat formation at different depths of the Earth.

The process begins with water heated at a great depth. The heat source may be hot rocks, partially molten or recently solidified but still hot magma associated with volcanic activity. The process could also occur along permeable fracture zones; the heating in this case would be due to deep and rapid circulation with regions of higher geothermal gradients. Hot water is less dense than cooler water and rises in the fracture networks as geothermal plumes. In some cases, the plumes leak to the surface through the caprock; however, most waters do not penetrate it. They slowly migrate away from the heat source where they cool, gain density, and sink back to the heat source to be reheated again and again to perpetuate geothermal convection. Large hydrothermal systems of this sort are stable and will last for thousands to a million years.

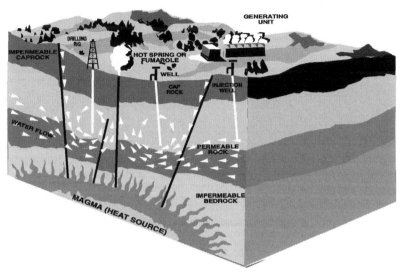

FIGURE 5.31 Graphic of hot rock technology.

Source: Courtesy of Energy Information Administration.

Surface effects of geothermal energy are only small fractions of the total available energy from the parent systems. Such systems may have volumes from tens to hundreds of cubic centimeters that are concealed. These can be detected and mapped using present-day geological, geophysical, and remote-sensing techniques. The Earth's core temperature increases with depth, at the center reaching greater than 4200 °C (7600 °F). Most of this heat is generated by the decay of radioactive isotopes with a small portion being a remnant of the globe's fiery formation 4.5 billion years ago. This heat convects to the surface naturally with an estimated 42 million thermal mW (42×1012 W) that continually radiates into space. Most of the heat cannot be used, however, because it reaches the surface at too low of a temperature.

The heat reaches the surface by a geologic process called *plate tectonics*. This process is also responsible for mountain building, volcanism, and earthquakes. Plate tectonics brings the Earth's concentrated heat to the surface at temperatures the can be used directly or for power generation. The process involves the floating of large plates of the Earth's surface on a bed of molten rock. The plates eventually collide; when two plates collide, one may be subducted (thrust beneath the other). At the great depth just above the downgoing plate, heat produces rock-melting temperatures. The resulting melting rock or magma is hotter and thus less dense than the surrounding rock, causing it to ascend through the upper mantle into the crust-forming volcanoes and large pools of heat.

Geothermal energy can be usefully extracted from four different types of geologic formations; hydrothermal, geopressurized, hot dry rock, and magma. Each of these reservoirs of geothermal energy can potentially be tapped and used for heating or electricity generation.

Hydrothermal reservoirs contain hot water and/or steam trapped in fractured or porous rock formations at shallow to moderate depths (100–4.5 km) by a layer of impermeable rock on top. The hot water and steam is formed either from an intrusion in the Earth's crust of molten magma from the interior, or the deep circulation of water through a fault or fisher. High-temperature hydrothermal resources from 180 °C to more than 350 °C are usually heated by molten rock. Lower temperature resources, in the 100 °C to 350 °C range, can be produced by either process. Hydrothermal reservoirs require a heat source, an aquifer with accessible water, and an impermeable rock cap to seal the aquifer. They have been the most common source of geothermal energy production worldwide. The reservoirs are tapped by drilling into the aquifer and extracting the hot water. **Geopressurized resources** are from formations where moderately high-temperature brines are trapped in a permeable layer of rock under high pressures at great depths. These brines often contain dissolved methane that can potentially be extracted for use as a fuel. The water and methane are trapped in sedimentary formations at a depth of about 3 to 6 km. The temperature of the water is in the range of 90 °C to 200 °C. Three forms of energy can be obtained from geopressurized reservoirs: thermal energy, hydraulic energy from the high pressure, and chemical energy from the burning of the dissolved methane. The major geopressurized reservoirs are located in the northern Gulf of Mexico. **Hot dry rock** is another potential geothermal resource. These reservoirs are generally hot impermeable rocks at depths shallow enough to be accessible (<3000 m). To extract heat from such formations, the rock must be fractured and a fluid circulation system developed (Figure 5.32). Although hot dry rock resources are virtually unlimited in magnitude around the world, only those at shallow depths are currently economical. **Magma**, the final, largest source of geothermal energy, is partially molten rock at very high temperatures (>600 °C). It is found at depths of 3 to 10 km and deeper. This great depth inhibits accessibility. Magma has a temperature that ranges from 700 °C to 1200 °C.

The use of geothermal energy can be divided into direct and indirect. Direct uses, in the form of geothermal springs, have been employed for heating, bathing, and cooking. Application of moderate temperature water (35–150 °C) has increased by 50% in the past five years. In the United States, the principal applications are fish farming; resorts and spas; greenhouses for growing vegetables, fruits, and flowers; and discrete and district heating of homes, workplaces, and other facilities.

US greenhouses today cover more than 110 acres, and domestic aquaculture annually yields an impressive 17,545,000 kg (38,600,000 pounds) of fish. Geothermal waters are also used in industrial applications, including heap leaching of precious metal ores, drying crops, and building materials. Geothermal heating systems are very durable. One system in Bosie, Idaho, has been operating continuously since 1892 with its two original production wells still in service.

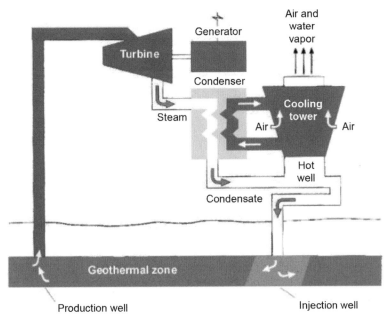

FIGURE 5.32 A dry steam power plant, closed cycle.

5.4.3 Geothermal electricity generation technologies

High-temperature geothermal energy can be harnessed to generate electricity. Currently, there is more than 8 GW of installed geothermal electrical capacity worldwide. The main conversion technologies used with a geothermal source are dry steam, flash steam, and binary cycle systems.

Dry steam power plants

A dry steam power plant is suitable where there is geothermal steam that is not mixed with water. Production wells are drilled down to the aquifer and the superheated pressurized steam (180–350 °C) is brought to the surface at high speeds, then passed through a steam turbine to generate electricity. In most cases, however, the steam is condensed back into water; it is called a closed cycle. This improves turbine efficiency and reduces environmental effects. The condensed water is injected back into the aquifer. The waste heat is vented through cooling towers. The energy-conversion efficiency is low, around 30%, similar to that of a fossil fuel plant.

The economics of a dry steam plant is affected by CO_2 and Hs. The pressure of these uncompressible gases reduces the efficiency of the turbine and the cost of operation is increased by their removal for environmental reasons. A closed-cycle dry steam power plant is shown in Figure 5.33. They are the simplest and most

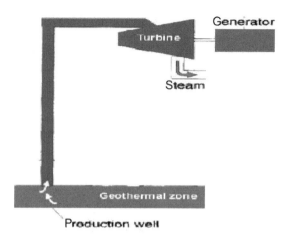

FIGURE 5.33 A dry steam power generation plant, open cycle.

economical and therefore widespread. The units for such technology are available in the 35 to 120 MW range. The United States and Italy have the largest dry steam resources. The Geysers field in California is the world's largest installed power source with a capacity of about 1100 MW.

Sometimes the dry steam drives the generator and then lets it go into the atmosphere; this type of electricity generating is called open-cycle dry steam power generation, as shown in Figure 5.34. This is simple but not efficient because the water is wasted.

Flash steam power plants

Flash steam power plants use single flash technology where the hydrothermal energy is in steam mixed with a liquid form (water), or just hot water. The fluid is sprayed into a flash tank, which is at a lower pressure than the mixture. Figure 5.35 shows a schematic diagram of a flash steam power plant.

Binary cycle power plant

A binary cycle power plant is used where geothermal resources are not sufficiently hot to produce steam, or where the source contains minerals or chemical impurities to allow flashing. In the binary cycle process, the geothermal liquid is passed through a heat exchanger. The secondary fluid with a lower boiling point than water (isobutane or pentane or amonia) is vaporized on the low-temperature side of the heat exchanger and expanded through the turbine to generate electricity. The working fluid is condensed and recycled for another cycle. All the geothermal fluid is injected back into the aquifer. A binary cycle power plant schematic is shown in Figure 5.36.

Binary cycle plants can achieve higher efficiencies than flash power systems, and they allow the utilization of lower temperature reservoirs. Other issues, such as corrosion and environmental concerns, are also eliminated. These systems are more expensive, and large pumps are required that consume a significant percentage of

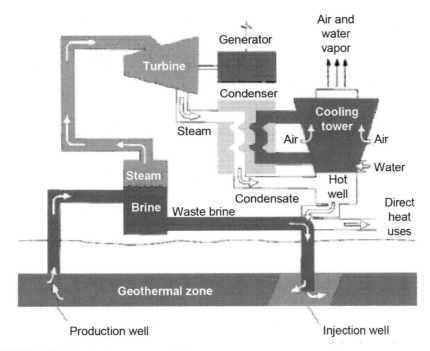

FIGURE 5.34 A flash steam power plant.

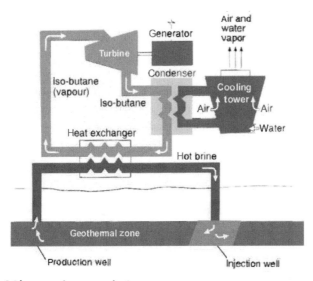

FIGURE 5.35 A binary cycle power plant.

FIGURE 5.36 A bus powered by biofuel (soybeans).

the plants' output. The units range in size from 1 to 3 MW. They generally are used in modular systems.

California generates the most geothermal electricity with about 824 MWe at The Geysers. Although this is much less than its capacity, it is still the world's most developed field. This is also the most successful renewable energy project in history. Other locations include 490 MWe in the Imperial Valley, 260 MWe at Coso, and 59 MWe smaller plants. California plans to add an additional 242 MWe. There are also power plants in Nevada (196 MWe with 205 MWe planned), Utah (31 MWe), and Hawaii (25 MWe). Other plants to total 75 MWe are planned countrywide.

Direct use has skyrocketed to 3858 GWh/yr, including 300,000 geothermal heat pumps. In the western United States, hundreds of buildings are heated individually and through district heating projects (e.g., Klamath Falls in Oregon, Bosie in Idaho, San Bernardino in California). Large greenhouses and aquaculture facilities in Arizona, Idaho, New Mexico, and Utah use low-temperature geothermal waters. Onions and garlic are also dried using this energy.

There are currently 12,000 MW of direct use and 8000 MWe of generating capacity in geothermal resources worldwide. This is about 0.4% of the world's total installed generating capacity. The United States, the Philippines, Italy, Mexico, Iceland, Indonesia, Japan, and New Zealand are the largest users of direct and indirect geothermal energy. Table 5.2 shows the location of present electric power generation in order of size by country. Note that the capacity in 1999 of 8246 MW was a 40% increase over the installed capacity of 1990. Other countries with less than 20 MW are Argentina, Australia, Ethiopia, France (Guadeloupe), Greece, Portugal (Azores), Russia, and Thailand.

5.5 BIOMASS ENERGY SYSTEMS

In contrast to the fuel based on products derived from the petrochemical industry, biofuel is based on a raw material derived from living organisms. The term "biomass" refers to organic matter that can be converted into energy. Some of the most common fuels for biomass energy systems (BES) are wood, agricultural residues,

Table 5.2 Countries Generating Geothermal Power in 2010		
Country	**Installed Capacity (MW)**	**Rank**
United States	3,086	1
Philippines	1,904	2
Indonesia	1,197	3
Mexico	958	4
Italy	843	5
New Zealand	628	6
Iceland	575	7
Japan	536	8
El Salvador	204	9
Kenya	167	10

and crops grown specifically for energy. In addition, it is possible to convert municipal waste, manure, or agricultural products into valuable fuels for transportation, industry, and even residential use.

5.5.1 Introduction

In a process called "photosynthesis," plants capture sunlight and transform it into chemical energy. This energy may then be converted into electricity, heat, or liquid fuels using a number of different energy-conversion processes. The organic resources used to produce energy using these processes are collectively called *biomass*. There are an uncountable number of woodstoves being used to produce heat for buildings or for cooking in the world, making biomass one of the most common forms of energy.

Utilities and commercial and industrial facilities are also using biomass to produce electricity because it is a versatile energy source. Other organic matter that is used as a source of biomass energy includes trees, timber waste, wood chips, corn, rice hulls, peanut shells, sugar canes, grass clippings, and leaves. Some animal wastes include cattle shed waste, poultry ether, sheep and goat dropping, slaughterhouse waste, and fisheries waste; municipal solid waste is made of human feces, urine, and household and yard residues.

Biomass is made up mainly of the elements carbon and hydrogen; technologies exist that free the energy from chemical compounds made up of these elements. However, biofuel can be burned directly and converted to usable energy forms. By burning the biomass, heat is generated, which in turn converts water to steam; then steam is channeled to a turbine that runs a generator to produce electricity. On the other hand, biomass can be converted into biofuel. There are two other types of conversions: biological and thermal.

The chemical composition of biomass varies among species; it consists of about 25% lignin and 75% carbohydrates or sugars. The carbohydrates' fraction consists of many sugar molecules linked together in long chains or polymers. Two larger car-bohydrate categories that have significant value are cellulose and hemicellulose. The lignin fraction consists of nonsugar-type molecules linked together in large two-dimensional, sheet-like structures that resemble "chicken wire." Nature uses the long cellulose polymers to build the fibers that give a plant its strength. The lignin fraction acts like a "glue" that holds the cellulose fibers together. It is this combination that gives plants their flexible resiliency to bend and sway like grasses and also the mas-sive structural strength to tower hundreds of feet like the ancient redwoods.

Through more effective forest management, short rotation forestry, advanced harvesting and processing techniques, and more efficient stoves and boilers, biomass can supply a large portion of the world's energy. In the United States, for instance, these techniques could increase the share of energy produced by biomass to more than 20% of US requirements. According to the World Bank, 50 to 60% of the energy in the developing countries of Asia, and 70 to 90% of the energy in the developing countries of Africa, comes from wood or biomass and half the world's population cooks with wood. Wood waste is used to fuel US utility power plants as large as 80 MW. Energy generation using wood has grown from 200 MW in 1980 to more than 7800 MW today.

5.5.2 Biomass energy conversion

Energy conversion is the transformation of energy from forms provided by nature to forms that can be used by humans. Over the centuries, a wide variety of devices and systems have been developed for this purpose. Some of the energy converters are quite simple. The early windmills, for example, transformed the kinetic energy of wind into mechanical energy for pumping water and grinding grain. Other energy-conversion systems are decidedly more complex, particularly those that take raw energy from fossil and nuclear fuels to generate electric power. Systems of this kind require multiple steps or processes in which energy undergoes a whole series of transformations through various intermediate forms.

Many of the energy converters widely used today involve the transformation of thermal energy into electric energy. The efficiency of such systems is, however, sub-ject to fundamental limitations, as dictated by the laws of thermodynamics and other scientific principles. In recent years, considerable attention has been devoted to cer-tain direct energy-conversion devices, notably solar and fuel cells that bypass the intermediate step of conversion to heat energy in electrical power generation. There are several ways of capturing the stored chemical energy in biomass as described in the following list.

Direct combustion: This is the burning of material by direct heat. Biomass, such as wood, garbage, manure, straw, and biogas, can be burned without processing to produce hot gases for heat or steam. This process ranges from burning wood in fireplaces to burning garbage in a fluidized boiler to produce heat or steam to generate electric power. Direct combustion is the simplest biomass technology and may be very economical if the biomass source is nearby.

Pyrolysis: This is the thermal degradation of biomass by heat in the absence of oxygen. Biomass feedstocks, such as wood or garbage, are heated to a temperature between 800 and 1400°F, but no oxygen is introduced to support combustion. The products of pyrolysis could come in three products: gas, fuel, and charcoal.

Anaerobic digestion: This process converts organic matter into a mixture of methane, the major component of natural gas, and carbon dioxide. Biomass, such as wastewater (sewage), manure, or food processing waste, is mixed with water and fed into a digester tank without air.

Alcohol fermentation: Fuel alcohol is produced by converting starch to sugar, fermenting the sugar to alcohol, then separating the alcohol–water mixture by distillation. Feedstocks, such as wheat, barley, potatoes, waste paper, sawdust, and straw, contain sugar, starch, or cellulose and can be converted to alcohol by fermentation with yeast. Ethanol, also called ethyl alcohol or grain alcohol, is the alcohol product of fermentation that is usable for various industrial purposes including alternative fuel for internal combustion engines.

Gasification: Biomass can be used to produce methane through heating or anaerobic digestion. Syngas, a mixture of carbon monoxide and hydrogen, can be derived from biomass. Hydrogen, expected to a very important fuel during the twenty-first century, is also obtained in this manner.

Landfill gas: This gas is generated by the decay (anaerobic digestion) of buried trash and garbage in landfills. When the organic waste decomposes, it generates a gas consisting of approximately 50% methane, the major component of natural gas.

Cogeneration: The process is the simultaneous production of more than one form of energy using a single fuel and facility. Furnaces, boilers, or engines fueled with biogas can cogenerate electricity for onsite use or sale. Biomass cogeneration has the potential to grow more than biomass generation alone because cogeneration produces both heat and electricity. Cogeneration results in net fuel use efficiencies of more than 60% compared to about 37% for simple combustion. Electric power generators can become cogenerators by using residual heat from electricity generation for process heat; however, waste heat recovery alone is not cogeneration

Biomass absorbs carbon dioxide during growth and emits it during combustion. Therefore, it "recycles" atmospheric carbon and does not add to the greenhouse effect. Low levels of sulfur and ash prevent biomass from contributing to the acid rain phenomenon. Nitrous oxide production can be controlled through modern biomass combustion techniques. Because of its low sulphur content, biomass can be cofired with coal in existing power plants to achieve compliance with laws such as the US Clean Air Act Amendments, which were enacted to abate air pollution primarily from industries and motor vehicles.

5.5.3 Bioenergy

The energy from the biomass (organic matter) is called bioenergy. The use of bioenergy has the potential to greatly reduce greenhouse gas emissions. Bioenergy generates about the same amount of carbon dioxide as fossil fuels, but every time a new plant grows, carbon dioxide is actually removed from the atmosphere. They require energy crops, such as fast-growing trees and grasses, called feedstocks. The following subsections describe three different bioenergy applications.

Biopower

Biopower is the use of biomass to generate electricity. There are four major types of systems: direct-fired, cofiring, gasification, and small modular.

Direct-fired combustion: Biomass is the second most utilized renewable power generation source in the United States. The direct-fired power plants are similar to the fossil fuel power plants. Around 7000 MW of power is produced by the biomass.

Cofiring: This involves replacing a portion of the coal with biomass at an existing power plant boiler, which represents one of the least costly renewable energy options.

Gasification: It is a thermochemical process that converts solid biomass raw materials into a clean biogas (methane) form. The biogas can be used for a wide range of energy production devices. For example, they are fuel cells and turbines; it also can be used as cooking gas.

Small modular systems: These biopower systems have a rated capacity of ~5 MW; potentially they can provide power at the village level to serve many of the people and their industrial enterprises. Around 2.5 billion people in the world are benefited by small modular biopower systems.

Biochemical

Using heat, the biomass can be chemically converted into fuel oil. The chemical conversion process is called pyrolysis, and it occurs when biomass is heated in the absence of oxygen. After pyrolysis, biomass turns into a liquid, called pyrolysis oil, that can be burned like petroleum to generate electricity.

Biofuels

Biofuels are fuels for direct combustion for producing direct electricity, but it is generally used as a liquid fuel for transportation (Figure 5.37). Woody and herbaceous plant tissue and animal wastes can be burned directly or converted into liquid or gaseous fuels. They can be used instead of, or together with, fossil fuels (e.g., coal, oil, and natural gas) to produce energy for electricity and for transportation and industrial uses.

Two common types of biofuels are used in United States: *bioethanol* and *biodiesel*. Bioethanol is from carbohydrates, and biodiesel is an ester made from fats or oils. Ethanol made from starch or sugar is contributing to renewable, domestic, and

FIGURE 5.37 McCommas Bluff's municipal solid waste landfill.

Source: Texas Comptroller of Public Accounts.

environmentally beneficial transportation fuels. Ethanol is made from cellulosic biomass such as herbaceous and woody plants, agricultural and forestry residues, and a large portion of municipal solid and industrial waste. Biofuels offer the United States and the world many benefits. They are good for the environment because they add fewer emissions to the atmosphere than petroleum fuels, and biofuels use wastes that currently have no use. Unlike petroleum, which is a nonrenewable natural resource, biofuels are a renewable and inexhaustible source of fuel.

Among trees suitable for energy crops are cottonwood, hybrid poplar, silver maple, black locust, sweetgum, eucalyptus, sycamore, and willow. Perennial herbaceous crops include bluestem, switchgrass, reed canary grass, wheat grass, some tropical grasses, and a few legumes; all these can be harvested like hay. Annual crops, such as corn, sorghum, and soybeans, already are being grown and processed for biofuels in some areas.

5.5.4 Types of biofuels

Ethanol

Bioethanol is an alcohol made by fermenting the sugar components of biomass. It is made mostly from sugar and starch crops. With advanced technology being developed by various research institutes, cellulosic biomass (e.g., trees and grasses) are also used as feedstocks for ethanol production. Ethanol can be used as a fuel for cars in its pure form, but it is usually used as a gasoline additive to increase octane and decrease vehicle emissions.

Biomass is the material that comes from plants. Through photosynthesis, plants use light energy to convert water and carbon dioxide into sugar that can be stored. Organic waste is also considered biomass because it began with plant matter. Some plants store energy as simple sugars, others store energy as complex sugars called a starch; another plant matter is cellulosic biomass, which is made up of very complex sugar polymers. This type of plant matter, which can be obtained from the following, is being considered as feedstock for producing bioethanol:

- Municipal solid waste (household garbage and paper products); see Figure 5.38
- Agricultural residues (leftover material from crops—stalks, leaves, and husks of plants; see Figure 5.39
- Energy crops (fast-growing trees and grasses) developed just for this purpose; see Figure 5.40

FIGURE 5.38 Agricultural residue: bagasse from sugar cane stalks (*left*) and rice husks (*right*).

FIGURE 5.39 Energy crops: (a) hybrid poplars, (b) switchgrass, and (c) willow tree species.

FIGURE 5.40 Forestry wastes: sawdust (*left*) and dead trees and branches (*right*).

FIGURE 5.41 Industrial wastes.

- Forestry wastes (sawdust and chips from lumber mills, dead trees, and tree branches); see Figure 5.41
- Food processing and other industrial waste (black liquor, a paper manufacturing byproduct); see Figure 5.42

Ethanol production

The following are the two main reactions for producing bioethanol from biomass:

Hydrolysis: This is a chemical reaction that converts complex polysaccharides in the raw feedstock to simple sugars. Acids and enzymes are used as catalysts in the process of converting biomass to bioethanol.

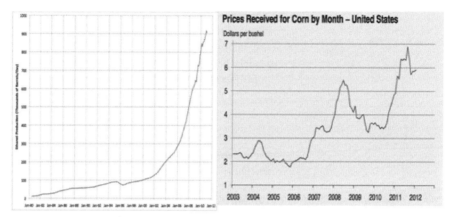

FIGURE 5.42 Increase in ethanol production and corn prices.

Fermentation: This is a series of chemical reactions that convert sugars to ethanol. Yeast and bacteria are used as catalysts in this chemical reaction. The fermentation reaction for the glucose is:

$$\underset{glucose}{C_6H_{12}O_6} \rightarrow \underset{ethanol}{2CH_3CH_2OH} + \underset{carbon\ dioxide}{2CO_2} \tag{5.9}$$

Future of ethanol

The future of ethanol is hopeful, but with certain limitations. According to the Renewable Fuels Association, 95 ethanol refineries produced more than 4.3 billion gallons of ethanol in 2005. An additional 40 new or expanded refineries that were scheduled to be opened by mid-2007 has increased the capacity to 6.3 billion gallons. This dramatic increase nonetheless represents just over 3% of our annual consumption of more than 200 billion gallons of gasoline and diesel. Figure 5.43 shows the projected increase in the production of ethanol and the subsequent increase in corn prices.

Moreover, one acre of corn can produce 300 gallons of ethanol per growing season. So, in order to replace that 200 billion gallons of petroleum products, US farmers would need to dedicate 675 million acres, or 71% of the nation's 938 million acres of farmland, to growing feedstock. Clearly, ethanol alone will not end dependence on fossil fuels. One possible solution is to replace most of our oil imports with food imports.

Biodiesel

Biodiesel is a clean-burning alternative fuel, which contains no petroleum, produced from domestic, renewable resources. It is typically produced from waste vegetable oil but can also be produced from a large variety of oils, new or waste, or from animal

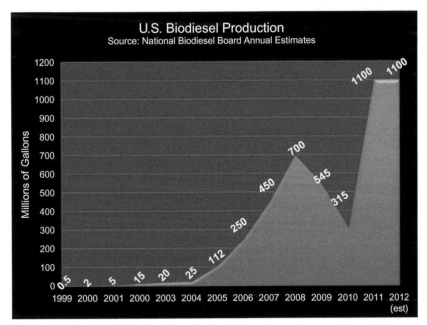

FIGURE 5.43 Estimated US biodiesel production by fiscal year.

fats. Biodiesel is a direct replacement for petrodiesel (petroleum-based diesel) that is made through a chemical process called "transesterification" whereby the glycerin is separated from the vegetable oil.

Using vegetable oil to power diesel engines is not a new concept. When Dr. Rudolf Diesel invented the diesel engine in the late 1800 s, it was originally designed to run on either fuel produced from petroleum or vegetable oil. When Dr. Diesel showcased his engine at the World Exhibition in Paris in 1900, he used 100% peanut oil for his demonstration. Dr. Diesel's feelings on diesel engine fuel made from vegetable oils are conveyed in two statements that he made. The first was made in 1911: "The diesel engine can be fed with vegetable oils *and would* help considerably in the development of agriculture of the countries which use it." His second statement in 1912 was: "The use of vegetable oils for engine fuels may seem insignificant today. But such oils may become in [the] course of time as important as petroleum and the coal tar products of the present time." Petroleum-based diesel fuel did not start to become popular until after Dr. Diesel's death in 1913.

Production of biodiesel

Transesterification. The transesterification process reacts in alcohol (e.g., methanol) with the triglyceride oils contained in vegetable oils, animal fats, or recycled greases, forming fatty acid alkyl esters (biodiesel) and glycerin. The reaction requires heat and a strong base catalyst such as sodium hydroxide or potassium hydroxide. The simplified transesterification process is:

$$\text{Triglycerides} + \text{Alcohol} \rightarrow \text{Fatty acid alkyl ester (Biodiesel)} + \text{Glycerin} \quad (5.10)$$

Once the reaction is complete, the major coproducts, biodiesel and glycerin, separate into two layers. The glycerin can be used in more than 1500 other products. For some vegetable oils containing more than 4% free fatty acids, an acid esterification process is used on the oil before the transesterification process begins to increase the biodiesel yield. Figure 5.44 shows the estimated US biodiesel production by fiscal year.

Supplying the oil and/or fats needed for biodiesel. Different crops yield different amounts of oil per acre. These values can range from 18 gallons per acre of corn grown up to 635 gallons per acre of oil palm grown. Knowing that there are 74 million surplus acres of US farmland, and assuming that each of the crops shown in Table 5.3 could grow in these surplus acres, the United States could supply anywhere from 1.3 to 47.0 billion gallons of vegetable oil, which corresponds to an equivalent amount of biodiesel. Since the United States consumes 30 billion gallons of diesel per year, there is a potential of having a 17 billion gallon surplus for export after covering the annual demand. This could also prevent the United States from importing the 277 billion gallons (6.6 billion barrels) of petroleum used to make petrodiesel.

These numbers are obviously not realistic unless oil palm is growable on the 74 million surplus acres of US farmland or additional farmland is used; however, it is a realistic possibility to significantly reduce or eliminate foreign oil dependence by producing biodiesel domestically. The table shows the amount of oil that can be produced for biodiesel, in gallons per acre, for a large variety of different crops. In the United States, soybeans are the crop of choice; however, in other parts of the world, rapeseed oil is the preferred crop.

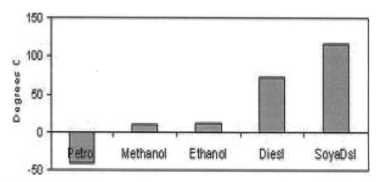

FIGURE 5.44 Fuel flashpoint comparison.

Table 5.3 Amount of oil produced from different crops

Crop	US gal/acre	Crop	US gal/acre
Corn (maize)	18	Safflower	83
Cashew nuts	19	Rice	88
Oats	23	Tung oil tree	100
Lupine	25	Sunflowers	102
Kenaf	29	Cocoa (cacao)	110
Calendula	33	Peanuts	113
Cotton	35	Opium poppy	124
Hemp	39	Rapeseed	127
Soybeans	48	Olives	129
Coffee	49	Castor beans	151
Linseed (flax)	51	Pecan nuts	191
Hazelnuts	51	Jojoba	194
Euphorbia	56	Jatropha	202
Pumpkin seeds	57	Macadamia nuts	240
Coriander	57	Brazil nuts	255
Mustard seeds	61	Avocado	282
Camelina	62	Coconut	287
Sesame seeds	74	Oil palm	635

Biodiesel uses and advantages

Biodiesel can be blended with conventional petroleum-based diesels, or used straight. When blended with petrodiesel, it is denoted as BXX, where XX is the percentage of biodiesel that is used (i.e., B20 is 20% biodiesel and 80% petrodiesel). Running biodiesel in its pure form will work in any diesel engine, though in nonbiodiesel-compatible engines, deterioration of some seals and hoses may occur over time. They can be replaced with biodiesel-compatible seals and hoses. Per their manufacturers, some automobile engines are 100% compatible with B100 or lower, while others are compatible with B5. The following are the advantages of biodiesel:

Safety: One of the biggest advantages of biodiesel is how safe it is for the environment and how safe it is to handle.
  - Nontoxic, ten times less toxic than table salt, making it safe to handle, transport, and store.
  - Less irritating to the skin than soap and water.
  - Biodegradable so it dissipates quickly, similar to dextrose sugar, if there were ever an oil spill. Also, since biodiesel is oil-based, it will not mix with seawater in the event of a tanker spill.
  - High flashpoint, 300 °F compared to 125 °F for petrodiesel, prevents it from easily igniting and increases the safety of handling it (Figure 5.45). In fact, "if you throw a match into a bucket of biodiesel, the match will go

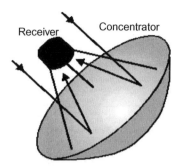

FIGURE 5.45 Schematic of a parabolic dish.

out," according to Leon Schumacher, associate professor of Agricultural Systems Management at the University of Missouri. He also states, "I've even pointed a propane torch directly at biodiesel. You wouldn't want to try that with petroleum diesel."

- First and only alternative fuel to have a complete evaluation of emission results and potential health effects submitted to the US Environmental Protection Agency (EPA) under the Clean Air Act's Section 211(b). (See Table 5.4.)

Emissions: They are substantially cleaner than petrodiesel emissions with no modifications to the engine or vehicle. Sulfur emissions are reduced 100% and CO_2, a greenhouse gas, is reduced 78% when compared to petrodiesel.

Fuel economy

Biodiesel and petrodiesel obtain comparable fuel economy when used in diesel engines. This fuel economy is 20 to 40% greater than the fuel economy achievable in gasoline-powered vehicles. An example of this is the Volkswagen New Beetle.

Table 5.4 Average Biodiesel Compared to Conventional Diesel
According to an EPA emissions study

Fuel Type	B100	B20
Regulated		
Total unburned hydrocarbons	−67%	−20%
Carbon monoxide	−48%	−12%
Particulate matter	−47%	−12%
NO_x	+10%	+2%
Unregulated		
Sulfates	−100%	−20%
Polycyclic aromatic hydrocarbons	−80%	−13%
nPAH (nitrated PAHs)	−90%	−50%
Ozone potential of speciated hydrocarbons	−50%	−10%

The 1.8-L gasoline engine has a fuel economy of 23 mpg in the city and 31 mpg on the highway, while the 1.9-L diesel engine has a fuel economy of 36 mpg in the city and 46 mpg on the highway. The fuel economy achieved with the use of diesel is comparable to the fuel economy achieved when using a hybrid car. Combining these two technologies is the most fuel-efficient option available. Prototypes of a diesel hybrid have achieved 90 mpg.

All renewable

All the ingredients required to produce biodiesel are renewable. The vegetable oil, which is the primary ingredient and was discussed earlier, can be obtained directly from a variety of crops. When the crop is harvested, it is pressed and the oil is extracted. The second main ingredient, methanol (CH_3OH), is produced during the fermenting of carbon-based compounds (e.g., wood) and is sometimes referred to as wood alcohol. The final ingredient, lye (KOH), can be made from wood ash and rainwater. Methanol can be purchased in a variety of locations and lye can be purchased from home and garden stores. All the energy required can be easily obtained with the use of solar power. The moderate temperature of 150 °F can be attained by solar heating, and an agitator (to ensure a well-mixed solution) can be run by a photovoltaic panel.

Disadvantages of biodiesel

Biodiesel has the following minimal disadvantages, most of which are easily remedied:

- Limited availability of refueling stations.
- Typically, it costs more than petrodiesel by $0.03 to $0.10 per gallon; however, with the recent spike in oil prices, biodiesel is actually $0.20 (or more) less expensive than petrodiesel.
- Really a disadvantage? Biodiesel can act as solvent and clean fuel lines of the buildup caused by petroleum diesel. This buildup can be released and clog fuel filters, resulting the need to change them routinely until the lines have been completely cleaned. This effect is also present with concrete-lined fuel tanks that had been used for the storage of petrodiesel, requiring them to be thoroughly cleaned before a conversion to biodiesel.
- Can soften and degrade certain types of elastomers and natural rubber compounds over time (i.e., fuel hoses and fuel pump seals). This problem can be remedied by replacing them with compatible elastomers. Some newer diesel engines are already 100% biodiesel compatible so that is not an issue.
- Will freeze (gel, really) around 25°F as opposed to 15°F for petrodiesel #2 (though other additives are typically added to petrodiesel in the winter to lower the freezing point). Can be remedied with an engine block heater, which most diesel engines already have.

Methanol

Methanol is methyl alcohol, commonly called wood alcohol. A common type of methanol used in cars is M85, a blend of 85% methanol and 15% gasoline. Methanol is predominantly produced by steam reforming of natural gas to create a synthesis gas. This gas is then fed into a reactor vessel in the presence of a catalyst to produce methanol and water vapor. Although a variety of feedstocks, instead of natural gas, can and have been used, the economy favors natural gas. Currently, all methanol produced in the United States uses methane derived from natural gas. However, methane also can be obtained from coal and biogas, which is generated by fermenting organic matter (e.g., byproducts of sewage and manure). Global attention has increased regarding producing methanol from biogas in an economically feasible way.

Advantages of methanol

Methanol's physical and chemical characteristics result in several inherent advantages as an automotive fuel. It is a potent fuel with an octane rating of 100, which allows for higher compression and greater efficiency than gasoline. Pure methanol is not volatile enough to start a cold engine easily, and when it is burning, it does so with a dangerously invisible flame. Blending gasoline with methanol to create M85 solves both these problems. Other benefits of methanol include lower emissions, higher performance, and lower risk of flammability than gasoline. In addition, methanol can be manufactured from a variety of feedstocks, most notably biomass. The use of methanol would help reduce US dependence on imported petroleum.

In addition, methanol can easily be made into hydrogen. Some researchers are currently working to overcome the barriers to using methanol as a hydrogen fuel source. This research may someday lead to methanol being used to create hydrogen for hydrogen fuel cell vehicles.

Disadvantages of methanol

Methanol is extremely corrosive, requiring special materials for delivery and storage. In addition, it has only 51% of the BTU content of gasoline by volume, which means its fuel economy is worse than that of ethanol. As with ethanol, any potential increase in efficiency from methanol's high octane is negated by the need for flex-fuel vehicles (FFVs) to remain operational on gasoline. Also, methanol produces a high amount of formaldehyde emissions. The lower energy content and the higher cost to build refineries for it compared with ethanol distilleries have made methanol a lower priority. Moreover, producing methanol from natural gas results in a net increase of CO_2, thereby hastening global warming. Unlike ethanol, the process liberates buried carbon that otherwise would not reach the atmosphere.

Future of methanol

The disadvantages seem to outweigh the advantages at this point. The EPA's Landfill Methane Outreach Program has been tasked with reducing methane emissions from landfills, and much of this methane is used to produce energy. As of December 2004, there were more than 325 operational landfill-gas energy projects in the United States and more than 600 landfills deemed to be good candidates for projects. Nevertheless, the quantities involved are relatively very small. Methane can also be produced by processing biomass such as grass clippings, sawdust, and other cellulosic sources.

Based on the important differences between ethanol and methanol, as well as the power of the farm lobby, ethanol is considered a higher priority as a replacement for gasoline. However, methanol may still have a future as a fuel. Nearly every major electronics manufacturer plans to release portable electronics powered by methanol fuel cells within the next two years.

5.5.5 Advantages and disadvantages of biomass

The following lists summarize the advantages and disadvantages of biomass in general.

Advantages
- Utilizes wastes.
- Versatile energy source. Organic matter used as a source of biomass energy includes trees, timber waste, wood chips, corn, rice hulls, peanut shells, sugar canes, grass clippings, leaves, manure, and municipal solid waste.
- Can be converted into biofuel and used in automobiles.
- Has a cleaning nature by getting rid of waste.
- Homegrown fuel that reduces dependence on foreign oil.

Disadvantages
- Could lead to deforestation.
- More expensive.

5.6 SOLAR THERMAL ENERGY CONVERSION SYSTEMS

Due to the depletion of fossil fuel deposits and the emission of greenhouse gases caused by industrialization and a higher worldwide demand for energy, there has been an urgent push to seek out alternative energy sources. One seriously studied solution is the utilization of solar rays for the production of energy. With global temperatures rising as a result of pollution, the world's high demand for energy can be constantly supplied by the semireliable cycle and temperature provided by solar rays using solar thermal energy conversion systems (STECS).

5.6.1 Introduction

Using solar rays for the production of energy is not always consistent and one major problem faced by researchers is how to deal with nighttime and cloudy days when solar rays aren't as available. The reality is that although the amount of sun that reaches the Earth is equal of that stored in coal, oil, and natural gas, solar power is more realistic in some areas of the world than others.

Solar thermal energy can be obtained by a STECS that exposes a collecting device to the rays of the Sun. Solar thermal systems use the warmth absorbed by the collector to heat water or another fluid or to make steam. Such systems can be classified in both small and large scale. Small-scale ones are typically for producing low- to medium-temperature heat to be used for space and water heating. Large-scale solar thermal systems are used to produce a low- to high-temperature heat for space heating and hot water for large buildings and agricultural and industrial processes. They can also be used to produce steam that can operate a turbine generator to produce electricity or industrial power.

5.6.2 Large-scale STECS

Most researchers of solar power systems for large-scale STECS are focusing on four technologies: (1) parabolic dish systems, (2) central receiver systems, (3) trough systems, and (4) solar pond systems. With the exclusion of the solar pond, large-scale solar thermal technologies realize high temperatures by using concentrators to reflect the rays of the Sun from a large area to a small receiver area.

Parabolic dish systems

As shown in Figure 5.46, parabolic dish systems use the concentrator to reflect sunlight to the receiver mounted on a focal point in front of a dish. The Sun is usually tracked using a computer-controlled, dual-axis parabolic concentrator. A transfer fluid absorbs heat as it travels through the receiver. In some systems, the heat engine, such as a Stirling engine, may be linked to the receiver to generate electricity. To achieve the higher efficiencies for converting solar energy, parabolic systems can reach higher temperatures of 1830 °F (1000 °C).

The power generation system, shown in Figure 5.46, captures the Sun's energy using a parabolic dish. The dish is lined with mirrors that focus the sunlight onto a receiver at the focal point. Liquid ammonia runs through the receiver where it is heated to 750 °C. The superheated ammonia gas runs a turbine and generator attached

FIGURE 5.46 Parabolic dish solar thermal system.

to the receiver spin at a speed related to the pressure of the ammonia gas (i.e., pressure changes with the temperature of the ammonia). This causes the frequency of the output to vary depending on the amount of solar energy. To compensate for this, the output of the generator is rectified to DC then changed back to AC at 60 Hz using an inverter.

Depending on the size of the dishes, they have an output power between 10 and 25 kW. A series of dishes can be deployed in a system to increase the total power of the plant. Parabolic dishes take up a relatively small space and are completely self-contained, making them an ideal candidate for remote location use. The dishes have the best solar-to-electricity efficiency, which is 12 to 25%, with peak efficiency measured at 29.4%. Power is stored in a battery backup.

Between 1982 and 1989, the Solar Total Energy Project (STEP) operated a large solar parabolic dish system in Shenandoah, Georgia (Figure 5.47). It included 114 dishes; each dish was 23 feet (7 m) in diameter. The STEP program produced high-pressure steam for electricity generation, medium-pressure steam for knitwear pressing, and low-pressure steam to run the air conditioning system for a nearby knitwear factory. The facility has been shutdown since October 1989 due to the failure of its main turbine and lack of funds for necessary plant repairs.

Parabolic dishes are considered a promising solar thermal technology because of their:

- Conversion efficiencies
- Modular flexibility
- Quick installation
- Minimal water requirements
- Siting flexibility

Disadvantages that delay the commercialization of the parabolic dish technology include:

- Lack of experience by companies using the technology
- Concerns that deployment of the technology in the distributed engine mode may suffer from excessive operation and maintenance costs
- Limited storage capacity

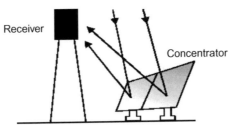

FIGURE 5.47 Schematic of a central receiver system.

FIGURE 5.48 A 10-MW central receiver solar energy conversion system.

Central receiver systems

The central receiver system consists of a field of thousands of mirrors, known as heliostats, surrounding a tower that holds a heat transfer fluid (Figure 5.48). Each heliostat has its own tracking mechanism to keep it focused on the tower to heat the transfer fluid. A typical system can achieve temperatures that range from 1000 to 2700 °F (538–1482 °C) at the receiver. A fluid gathers heat as it circulates through the receiver and transports it to a thermal storage subcomponent, which in turn supplies heat directly for industrial applications, or operates a turbine to generate electricity.

Much of the current research at Sandia National Laboratories, a test facility near Albuquerque, New Mexico, is on central receiver systems funded by the US Department of Energy (DOE). Sandia is testing molten salt transfer fluids and improved heliostats. The molten salt transfer fluid technology and the improved heliostats were used until 1999 in the Solar Two project in California, a 10-MW system (Figure 5.49).

According to the DOE, during daylight hours, 2000 mirrors at Solar Two tracked the Sun and stored its energy as heat in molten salt. This energy was then used to generate electricity when needed such as during periods of peak demand for power. The advantages of central receiver technology include:

- The ability to store energy and offer dispatchable power that is important to energy providers.
- High conversion rates and potentially lower capital outlay and operating costs due to the piping and plumbing of the system.

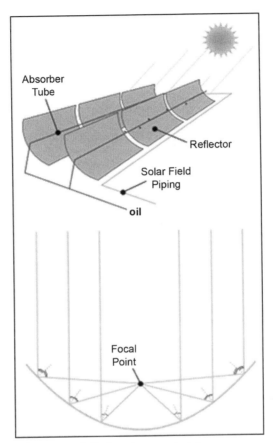

FIGURE 5.49 Schematic of a parabolic trough concentrator.

Trough systems

These solar concentrators are curved in only one dimension, forming troughs. The troughs are mounted on a single-axis tracking system that follows the Sun from East to West. They are lined with a reflective surface that focuses the Sun's energy onto a pipe located along the trough's focal line (Figure 5.50). A series of parabolic trough concentrators is used to construct a solar thermal trough system.

The parabolic trough system shown in Figure 5.51 includes long rows of trough concentrators. A heat transfer fluid is circulated through the pipes and then pumped to a central storage area, where it passes through a heat exchanger. The heat is then transferred to a working fluid, usually water, which is flashed into steam to drive a conventional steam turbine. Parabolic trough condensers liquefy the steam exiting the turbine. Receiver temperatures can reach 750 °F. A plant is typically supplemented with a natural gas-fired superheater to produce steam for utility-scale applications. A cooling tower used to cool the water utilized to condense the low pressure steam coming out of the turbine.

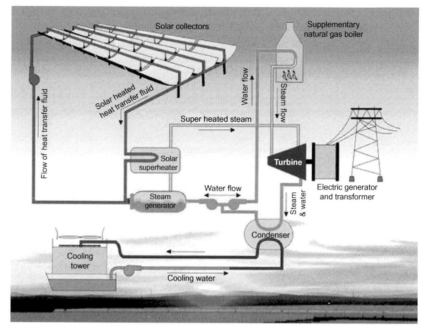

FIGURE 5.50 Luz International's solar thermal trough system.

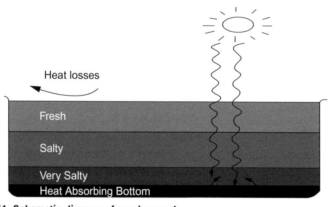

FIGURE 5.51 Schematic diagram of a solar pond.

Luz International installed the 13.8-MW Solar Electric Generating System I (SEGS I) in Daggett, California, in 1984. Oil is heated in the receiver tubes to 650 °F (343 °C) to supply steam to turbines for electricity generation. SEGS I contains six hours of thermal storage and uses natural gas-fueled superheaters to supplement the solar energy. Luz also constructed additional plants, SEGS II through VII, each with a 30-MW capacity. In 1990, Luz completed construction of SEGS VIII and IX at Harper Dry Lake in California, each with a 80-MW capacity. The total generating capacity of these plants is 350 MW, individually and combined, which makes them the largest solar power plants in the world.

The biggest advantage that parabolic troughs have over the other STECS technologies is their relatively advanced stage of commercialization. The greatest disadvantages of solar parabolic trough systems include low conversion efficiency when compared with other solar thermal technologies, the need for a supplemental fuel source used primarily to prevent the heat transfer fluid from freezing, and insignificant amounts of water needed for cooling. Even though dry cooling is possible, it results in performance degradation and an increase in initial capital costs.

Solar ponds

A solar pond is a body of water used to collect and store solar energy. The pond can be either natural or human-made, and it contains salt water, which acts different from fresh water. In a freshwater pond, sunrays entering the pond would heat the water and, by natural convection, the heated water would rise to the top, while the heavier cooler water would sink to the bottom. Salt water, on the other hand, is heavier than fresh water as shown in Figure 5.52 and will not rise or mix by natural convection. This creates a larger temperature gradient within the pond. Fresh water forms a thin insulating surface layer at the top, and underneath it is the salt water that becomes hotter with depth—as hot a 93°C at the bottom.

The most common use for solar salt water gradient ponds is the generation of electricity. Heated brine is drawn from the bottom of the pond and piped into a heat exchanger, where its heat converts a liquid refrigerant into a pressurized vapor that spins a turbine, generating electricity. Solar pond desalination systems can also be used to provide drinking water in areas where fresh water is scarce.

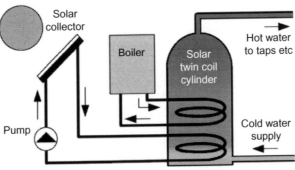

FIGURE 5.52 Principles of heating water for a household using solar thermal energy.

5.6.3 Small-scale STECS
Residential hot water
Installing solar hot water heaters in residential homes can be economical; a solar system can satisfy 60 to 80% of hot water needs. Standard tanks for consumer use in the United States range from 60 to 120 gallons, and homes can often be easily updated with a solar system. Most homes with solar thermal systems have a backup water heating system in case of cloudy weather or extended hot water use. Figure 5.53 shows how thermal collectors function with flat plate collectors that use the Sun's rays to heat a circulating fluid, which in turn provides usable heat to a household. The fluid, in this case water, flows through copper tubing in the solar collector and in the process absorbs some of the Sun's energy. The circulating fluid then moves to the heat exchanger, where it warms the water used by the household. Finally, a pump moves the fluid back to the solar collector to repeat the cycle.

Commercial and industrial hot water
Solar thermal systems provide hot water or space heating to about 250,000 commercial and industrial buildings in the United States. Such systems are most commonly found in the laundry, food service, food processing, metal plating, and textiles industries.

Heating swimming pools
Simple solar collector systems can be used to raise the temperature in swimming pools by 8 to 10 °F in order to extend the swimming season three or four months. Pool systems differ from residential hot water ones in that the pool serves as a storage tank; a pump is used to circulate water through the system and the collector is black plastic tubing with no glazing.

FIGURE 5.53 A solar cooker used in India.

Solar cooker

Conventional fuels can be saved by supplementing them with energy by using a solar cooker to cook food. It supplements the cooking fuel but cannot replace it entirely. Solar energy is abundantly available in India, therefore, on a clear, sunny day, it is possible to cook lunch for 4 to 5 people in a normal box solar cooker (Figure 5.54). Additionally, full or part of the evening meal could be cooked in it. More than 487,000 solar cookers have been sold in India since the Ministry of Non-conventional Energy Sources has been promoting the sale of them. Nearly 40 manufacturers are involved in the fabrication of solar cookers.

Sunspaces

Greenhouses or sunrooms that are an extension of a building is a habitable solar collector and is common in cooler climates (e.g., southern and eastern Australia and New Zealand). Heat is generated in the sunroom using incoming solar radiation; this in turn preheats the air before it circulates into the main building. Heat energy is stored within the thermal mass of the building, especially the wall between the sunroom or conservatory and the remainder of the house (Figure 5.55).

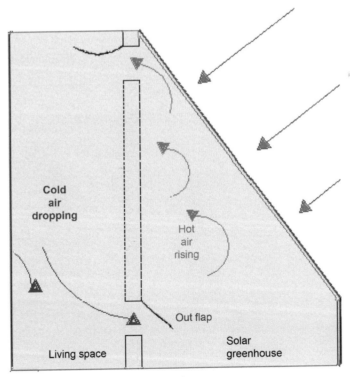

FIGURE 5.54 Schematic of a sunroom conservatory.

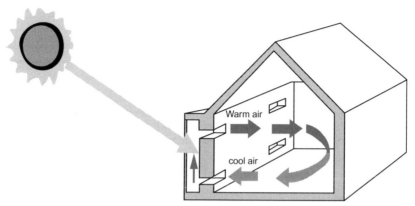

FIGURE 5.55 Schematic of a Trombe wall.

Trombe wall

The Trombe wall builds on the idea of utilizing a space in the building to conserve energy. This was developed by Felix Trombe in the 1950s and was first used in the south of France. In this space-heating system, a thin air space is behind a storage wall (used for storing heat energy); solar radiation is absorbed and stored as heat in the thermal mass contained within it. Heat is then circulated by radiating it directly into the building (Figure 5.56).

Windows between the Trombe wall and the building can be opened to allow circulation of the preheated air, which increases the heating performance of the wall.

FIGURE 5.56 Photo of three microturbines.

The windows can be closed to prevent the loss of energy at night or on cold days. There are many varieties of the Trombe wall, particularly in situations where modifications are being made to an existing structure; for example, small-scale collector panels or boxes fitted to the roof or wall, flat plate collectors, or the covering of the Sun-facing wall (North) with transparent insulation.

5.6.4 Advantages and disadvantages of STECS

According to the Energy, Resources, & Technology Division of the Hawaii Department of Business, Economic Development & Tourism, the following are the advantages and disadvantages of using solar thermal energy:

Advantages
- Use of a renewable natural resource.
- Pollution-free energy.
- Ability to store energy for use during cloudy or overcast periods.
- Competitive cost to fossil fuel energy due to new advances in technology.
- Reduced dependence on imported fossil fuels.
- Tax incentives; for example, the State of Hawaii offers a 35% tax credit for the cost of buying and installing a solar thermal device that supplies hot water for single-family homes or multiunit residential buildings.
- Readily available; simple to use, install, and maintain equipment.

Disadvantages
- Not economical in areas that have long periods of cloudy weather or short daylight hours. Efficiency is also reduced by atmospheric haze or dust.
- Freezing can damage collecting system components such as pipes.
- Large collecting devices may use extensive areas of land that may be needed for agricultural, residential, resort, or commercial purposes.
- Cosmetically, the reflective devices may be disturbing or undesirable in certain areas.
- These systems do not operate at night or in inclement weather.
- Storage of fluids at the higher temperatures needed for electrical generation, or storage of electricity itself, needs further research and development.

Given the many advantages to using STECS, there has been much research in this area, most of it focuses on the four types of large-scale solar thermal energy systems.

5.7 MICROTURBINES

Microturbines are among the newest form of energy generation available for the market. If used properly, they are efficient and dependable.

5.7.1 **Definition and applications of a microturbine**

Microturbines are a new type of combustion turbine being used for stationery energy generation applications. They are small combustion turbines, approximately the size of a refrigerator, with outputs of 25 to 500 kW. They can be located on sites with space limitations for power production. Waste heat recovery can be used in combined heat and power systems to achieve energy efficiency levels greater than 80%. In addition to power generation, microturbines offer an efficient and clean solution to direct mechanical drive markets such as compression and air conditioning. Best of all it is relatively small; a typical commercial 60-kW microturbine is 0.76 m wide × 1.93 m long × 2.08 m high. The five manufacturers are Capstone, GE, Ingersoll Rand, Solar Turbines Incorporated, and United Technologies. Figure 5.57 shows three microturbines side by side, with a man in the picture for scale.

"Microturbine and engine generator products are electricity-producing assemblies typically located at or near the point of use. They are generally installed so that backup (standby or emergency) power is available to the user in the event of a utility grid failure. In some installations, these units can be connected in parallel with the local electric utility power grid and used for peak sharing, or excess power can to sold back to the utility. In other applications, microturbines and engine generators are located off the grid in rural and remote areas where they provide the sole source of power (prime or continuous), or operate in combination with other sources such as photovoltaic or wind turbine installations.

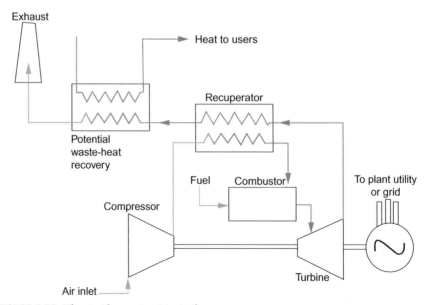

FIGURE 5.57 Microturbine system block diagram.

Internal combustion, gasoline, natural gas, propane, and diesel engines from the automotive and marine industries have been coupled to rotating field alternator and generator devices for many years. In addition, the turbine engines, now being coupled to high-speed generators in microturbine applications, are from the same family of small jet engines that have been used in the military and transportation industries for the past 50 years. These systems are considered to employ the most reliable power-producing technologies ever used in stand-alone and distributed systems. Typically, microturbines are kept running at all times because of there main drawback—the limited number of times they can be started up and shutdown. So, the most efficient use is in the off-grid situation.

5.7.2 Components and operation of microturbines

Microturbines are composed of a compressor, combustor, turbine, alternator, recuperator, and generator. Microturbines can also be classified as simple-cycle or recuperated. In a simple-cycle, or unremunerated, turbine, compressed air is mixed with fuel and burned under constant pressure conditions. The resulting hot gas is allowed to expand through a turbine to perform work. Simple-cycle microturbines have a lower cost, higher reliability, and more heat available for cogeneration applications than recuperated units. Recuperated units have a higher thermal-to-electric ratio than unrecuperated units and can produce 30 to 40% fuel savings from preheating.

Figure 5.58 shows the microturbine system. Its efficiency without heat recovery is only ~25.30%, but with heat recovery it is 70 to 80%. Heat recovery is a system that takes the exhaust gas from the turbine then passes it through the heat recuperator,

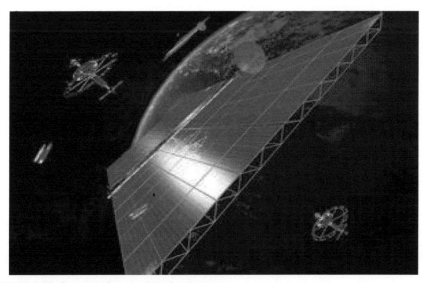

FIGURE 5.58 Concept of space-based solar power.

which takes the heated air and heats the compressed air before it enters the combustion chamber. Preheating reduces fuel usage, thereby increasing efficiency. Microturbines tend to be highly reliable; they require nearly no maintenance because they have almost no moving parts.

5.7.3 Advantages and disadvataages of microturbines

The following lists summarize the advantages and disadvantages of microturbines in general.

Advantages
- *High efficiency:* See earlier discussion for efficiency details.
- *Durability:* Can be used for 11,000 hours of reliable operation between major overhauls, with a service life of at least 45,000 hours.
- *Fuel flexibility:* Options for using multiple fuels include diesel, ethanol, landfill gas, and biofuels. In today's market, fuel flexibility is extremely important. As you already know, gas prices can be as much as $4.19 per gallon, sometimes more. If this system only ran on gas, there would be no market for it.
- *Clean emission system:* The major environmental benefits include the ultralow-emission exhaust that is oxygen-rich and contains less than nine parts per million of nitrogen oxides. Capstone's microturbines can use methane "waste" gas produced at sewage treatment plants, agricultural digesters, and landfills to generate power with near zero emissions.
- *Work on diversified fuel:* Natural gas, propane, diesel, and kerosene can fuel microturbines. They also can use oil field "waste" gas as fuel, effectively destroying pollutants and greenhouse gases and creating onsite electricity in remote locations where there is no electric grid.

Because of their compact size and low operation and maintenance costs, microturbines are expected to capture a significant share of the distributed generation market.

Disadvantage
- *High initial cost per kW*: In 2003, the units were priced at $\sim$\$250,000, which no doubt has increased. This price included a fuel booster, a compressor, a heat recovery module, application engineering/commissioning, and first year maintenance. Installation was $\sim$\$100,000. The hardware cost is $\sim$\$1100/kW, including heat exchanger and gas fuel compressor.

5.8 SATELLITE POWER SYSTEMS

Some estimates show that by 2050 more than 10 billion people will inhabit the globe, more than 85% of them in developing countries. The question is: How we can supply humanity's growing energy needs without the adverse impact on the environment?

Dependence on fossil fuels is not the answer because burning coal, oil, and gas will pour carbon dioxide into the atmosphere, raising the risk of further global climate change. Of course, current resources will not last forever. Nuclear energy avoids the greenhouse problem but introduces the problem of disposing of nuclear waste.

Terrestrial renewable sources, such as hydropower, solar, biomass, terrestrial photovoltaic, biomass fuels, and windfarms, generate energy; however, the flow of power from them is intermittent—clouds blot out the Sun, the wind stops blowing, lack of rainfall nulls hydropower generation. Because these technologies do not deliver power continuously, they require some means of storing energy, which adds to the overall cost and complexity. A network of solar-powered satellites in a low Earth orbit could provide power to any location on a continuous basis because at least one satellite would always be in the receiving station. The solar power satellite (SPS) concept uses sunlight in space to generate baseload electricity on Earth.

5.8.1 Introduction

Orbiting satellites would collect solar energy and beam it to Earth where it would be converted to electricity (Figure 5.59). Several different methods are possible, including microwave, laser, and mirror transmission; however, the one that has received the most effort is the use of microwave beams or wireless power transmission. This method is good because sunlight diffuses and is not available continuously at the Earth's surface. An additional possibility is to collect solar energy 24 hours a day in space and transmit it to Earth using a solar power satellite. An SPS has

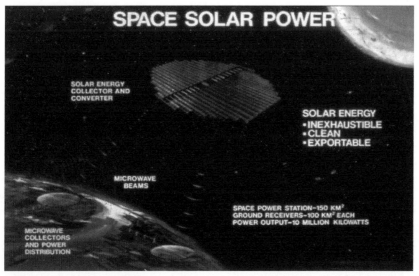

FIGURE 5.59 A microwave SPS.

the potential to supply several hundred gW of baseload electrical power. The need for SPS system and its development will depend on a very high increase in demand for electricity (i.e., if coal and nuclear options are constrained).

Solar power satellites were invented by a Czech-American, Dr. Peter Glaser of Arthur D. Little, in 1968. Following several years of preliminary studies, and driven by the impetus of the oil crises of the time, a major study of power from space was conducted by the then newly created Department of Energy with the assistance of NASA. Instead promoting the concept, the results of the 1970s study led to the stoppage of any serious consideration of US space solar power. The United States wanted international funding to continue the research, money that didn't come; this was added to the fact that the energy lobbyist did not want to move in that direction.

5.8.2 Types of solar power satellite

The SPS concept is directed toward providing baseload electrical power by using the photovoltaic cell principle to convert the Sun's energy directly into electrical power. The concept seeks to remove the harsh limitations placed on the terrestrial PV arrays for baseload electrical power during day–night time and the variability of the intensity of solar energy caused by changing atmospheric conditions. The SPS needs several technological improvements, especially in the fields of transportation vehicles and space stations needed to construct the orbiting solar power plant. Also, questions of environmental, economic, societal, and political feasibility must be answered. Therefore, if this technology becomes a reality, it probably will not be for many years. There are three types of SPS transmission, as described in the next subsections.

Microwave transmission

Solar energy would be collected and converted into microwaves. These microwaves would be beamed to a receiving antenna on Earth where it would be converted to electricity. The reference system satellite design consists of a 55-km^2 flat array of photovoltaic solar cells located in an orbit 35,800 km above the equator. The cells convert solar energy into direct current that is conducted to a 1 km diameter microwave transmitting antenna mounted at the end of the PV array.

Microwave transmitting tubes (klystrons) convert the current to radio frequency power at 2.45 GHz and transmit it to Earth. A ground antenna (rectenna) receives the electromagnetic radiation and rectifies it back to direct current (Figure 5.60). The DC power can be inverted to AC power and stepped up to high voltage. It would be either rectified to DC and delivered directly to a DC transmission network in the terrestrial utility grid or used as conventional AC power. The rectenna on Earth covers a ground area of 102 km^2 and would require an extra area of 72 km^2 to protect against exposure to low-level microwaves, and the beam density at the center of this antenna is 23 mW/cm^2.

FIGURE 5.60 Photo of a receiving station (rectenna).

Laser transmission

Solar radiation would be collected and converted to infrared lasers, which would be beamed to an Earth receiver. Compared with microwaves, lasers have a much smaller beam diameter; since the area of both transmitting and receiving antennas decreases as the square of the wavelength, light from an infrared wavelength laser can be transmitted and received by apertures more than 100 times smaller in diameter than a microwave beam. The great reduction in aperture area allows, for example:

- Use of low sun-synchronous rather than geostationary orbits. (This orbit is a near-polar low Earth orbit that keeps the satellite in full sunlight all the time.)
- The primary laser would beam its power up to low-mass laser mirror relays in the orbit for reflection down to the Earth receiver. This would reduce the cost of transportation, since the system will be in low Earth orbit rather than in geostationary orbit.
- A laser system might be able to operate efficiently and economically on the smaller scale (100–1000 MW). This offers the flexibility of power demand matching on the ground.
- Laser power transmission would avoid the problem of microwaves' biological effects and reduce overall interference with other users of the electromagnetic spectrum.

A laser SPS has some disadvantages:

- *Absorption of laser radiation*: Infrared radiation is subject to severe absorption by clouds. A baseload system, unlike microwaves, would require considerable storage capacity to make up for interruptions—these are expensive.

- *Efficiency*: Current high-power, continuous-wave lasers are only capable of very low overall power conversion efficiencies (<25%). Converting the beam back into electricity is also inefficient.
- *Health and safety hazard*: The beam's intensity would be great enough to constitute a hazard.

Mirror transmission

Orbiting mirrors would reflect sunlight to a central station on the Earth. The receiver would convert 24 hours of illumination into electricity. Instead of placing an energy-conversion system in orbit, large orbiting mirrors could be used to reflect sunlight to ground-based solar conversion systems. Thus, the system's space segment could be much simpler, therefore less expensive and more reliable.

One such system would consist of a number of circular plane mirrors in Earth orbits, each of which directs sunlight to the collectors of ground-based solar–electric power plants. Conversion from sunlight to electricity would occur on the surface of the Earth. In one approach (the "SOLARES baseline" concept), about 916 mirrors, each 50 km^2 in area, would be required to produce 810 GW from six individual sites. A number of different mirror sizes, orbits, and ground station sizes are possible.

A more feasible option would be a lower orbit system (2100 km) to supply 10 to 13 GW per terrestrial site. One of the features is that it could be used for either solar–thermal or solar–PV terrestrial plants. The fact that energy conversion would take place on the surface of the Earth keeps the mass orbit small, therefore reducing transportation cost. However, a major disadvantage of such a mirror system would be that the entire system would require an extremely large land area for the terrestrial segment, although placement in populated locations could be avoided by using arid land and desert areas.

5.8.3 Advantages and disadvatages of a SPS

The following lists summarize the advantages and disadvantages of solar power satellite system in general.

Advantages
- *Efficiency*: The SPS system is very efficient.
- *High availability:* The system would have an availability of 99%, the amount of time that the Sun is illuminated in space. The reason for its high availability is the high reliability due to the long life of the components that will be used and the absence of any rotating or reciprocating motions in which mechanical friction is used. Also, much redundancy is built into the system. Unpredicted outages should be caused only by extremely severe weather at the rectenna site or a meteor crashing into a sensitive part of the satellite, both of which have very low probabilities.
- *Availability to any location*: The energy can be beamed to any spot on Earth that needs the energy, especially in emergencies.
- Promotes cooperation among nations.

Disadvatages

- Radio frequency interference from random noise and harmonics of the transmission frequency (2.45 GHz).
- Local heating of the ionosphere, which could interfere with communications links such as long-range and satellite communications passing through the affected area.
- *Possible harmful effects:* The US Energy and Commerce Committee has stated that the reference system frequency of 2.45 GHz, and its second through fifth harmonics, could interfere with communications, radar, navigation, broadcast satellites, computers, and some medical devices and instruments. The RF interference and ionospheric problems are being studied by the US DOE. Safety features, such as a pilot beam for the transmitting antenna to track, are planned to keep the microwave beams from wandering off target and affecting people. An exclusion area surrounding each rectenna is proposed to hold the risk to humans near the target site to a minimum.
- *High cost:* Current overall cost estimates for the SPS and its major components are highly uncertain. The assessments of upfront costs range from $40 to $100 billion. The most detailed estimate was made by NASA for the reference design. These call for a 22-year investment of $102.4 billion (including transportation and factory investment) to produce the first 5-GW satellites, with each additional satellite costing $11.3 billion. The total costs estimated by NASA include major elements such as space transportation and PV cells.

The cost is a major obstacle in realizing this technology. The SPS system has been called the most technically challenging project to be considered seriously since the beginning of the space program. The cost of constructing one satellite and ground station is estimated at $12 billion. The total system cost to supply 300 GW is put at three-quarters of a trillion dollars. Thus, the estimates could eventually turn out to be high or low. By current estimates, capital cost would be in the range of $2000 to $4000/kW generated. This translates into electricity that costs $05.5/kWh at the rectenna (Figure 5.60) before it is transported to consumers. These projections indicate that the solar power satellite would be competitive with other power-generation alternatives.

5.8.4 The future of SPS

Using the technological approaches that were at hand during that era, the DOE–NASA study created a "1979 Reference SPS System" design for solar power satellites that quickly became the focus of discussion and debate. The SPS system is thought to be the future for energy supply around the world. Every region could have energy from a satellite. It may sound like a fantasy, but in a few years it may not be a dream anymore."

For that to happen, however, a global effort has to be initiated. Countries have to come together and put the money on the table. It will be a very expensive endeavor; the technology is there and also the human capital. The first study on the SPS by NASA was in 1968 as a result of an oil and gas crisis. Some 45 years later, the budget for that research has been completely diminished. One of the reasons is that big oil companies are against an SPS system and are doing an excellent job of lobbying lawmakers.

Today there is a new incentive in favor of implementing a solar power satellite system; it is another source of "green energy" to reduce global warming because the Earth is getting hotter and hotter. It is recognized that oil and gas are the main sources of the pollution that causes warming. Governments around the world are taking steps to limit the use of both. It now might be easier to "sell" the SPS project to countries and governments as an alternative to fossil fuels.

Many states are starting to think about green energy because incentives are being given to companies that are working on reducing pollution. Government funding is available for those interested in PV and other alternatives. But, to implement the SPS system, it will take more than one country and more than incentives. It will take commitment and big money from many countries around the world.

The satellite power system, in the form of a solar power satellite system, is very simple in its conceptual design; however, the practical feasibility of one will require a lot of regulation and work. The concept is to put a PV system on a satellite placed in the geosynchronous orbit so that it remains above a fixed point on the Earth's equator. The PV will collect the Sun's radiation and produce electricity, which will be sent to Earth. An alternative concept envisions using large orbiting reflectors to reflect solar radiation to the ground, creating immense solar farms where sunlight would be available around the clock. We can also use laser beams as a transmission medium to carry the energy.

PROBLEMS

5.1   The basin of a tidal power station in Shepody, New Brunswick, has an area of 115 km^2. The tidal range is 10 m, and the salt water density is 1025.18 kg/m^3. Calculate the power output.

5.2   From a tidal power plant in Puerto Rio Gallegos in Argentina, the tidal range is 14.5 m, the surface area of the basin is 9×10^5 m^2, and the efficiency of the generator turbine is 60%. Calculate the power output of the generator if the salt water's density is 1025.18 kg/m^3.

5.3   For a wave in deep water having a time period of 5 sec and a height of 0.91 m:
(a) Calculate the wavelength.
(b) Calculate the phase velocity.
(c) Calculate the total potential energy.
(d) Calculate the power output.

5.4   Assume we have the following wave amplitude, $a = 12$ (m), and the wave length is $\lambda = 5$ (m):

    (a) Calculate the period, T.

    (b) Calculate the power output.

5.5   The flooded area of a basin is 10 km square, the variation in height between low and high tide is $R = 2$ m, and the tidal period is 12.4 hours. Calculate the maximum power available from such tidal system.

5.6   What is the theoretical and real efficiency of an OTEC if the surface ocean temperature is 75 °F and the deep water temperature is 55 °F?

5.7   For problem 5.7, calculate the power output of the OTEC system if the flow rate of both the warm and cold water is 5000 kg/sec and the turbine generator efficiency is 80%.

References

[1] *www.green-trust.org/otec.htm; [accessed 14.10.12].

[2] www.celsias.com/article/can-ocean-thermal-energy-conversion-turn-tide-clim/; [accessed 14.10.12].

[3] http://newenergyportal.files.wordpress.com/2009/12/closed-cycle-ocean-thermal-energy-conversion-otec1.png; [accessed 14.10.12].

[4] *www.oceansatlas.org/unatlas/uses/EnergyResources/Background/OTEC/OTEC2.html.

[5] *www.otecafrica.org/Project.aspx; [accessed 14.10.12].

[6] Nova Scotia Power. Annapolis tidal generating station brochure; July 2001. www.worldenergy.org/wec-geis/publications/reports/ser/tide/tide.asp(10.14.12).

[7] *www.esru.strath.ac.uk/EandE/Web_sites/01-02/RE_info/Tidal%20power%20files/image002.jpg [10.14.12].

[8] UEK Ltd. Underwater electric kite.

[9] *www.bluenergy.com/tidal.html.

[10] www.pol.ac.uk/home/insight/tidefaq.html; [accessed 14.10.12].

[11] *www.bwea.com/marine/devices.html; [accessed 14.10.12].

[12] *http://inventors.about.com/od/tstartinventions/a/tidal_power.htm; [accessed 14.10.12].

[13] *http://webecoist.com/2008/11/09/hydroelectric-wave-tidal-power/; [accessed 14.10.12].

[14] *http://en.wikipedia.org/wiki/Wave_power; [accessed 14.10.12].

[15] www.eia.doe.gov/kids/energyfacts/sources/renewable/ocean.html; [accessed 14.10.12].

[16] *Power Plant Specifications, US Department of Energy Geothermal Technical Website. http://en.wikipedia.org/wiki/Geothermal_energy; [accessed 14.10.12].

[17] *NREL's Geothermal Technologies Program Website. *www.nrel.gov/clean_energy/geoelectricity.html.

[18] www.geo-energy.org/publications/reports/Geo101_Final_Feb_15.pdf.

[19] *http://egi-geothermal.org/GeothermalEnergy/ElectricPower.htm; [accessed 14.10.12].

[20] *http://geothermal.id.doe.gov/; [accessed 14.10.12].

[21] *http://geothermal.marin.org/GEOpresentation/sld059.htm; [accessed 14.10.12].

[22] *www.nevadageothermal.com/i/pdf/What-is-geothermal-power.pdf; [accessed 14.10.12].

[23] https://inlportal.inl.gov/portal/server.pt?open=512&objID=422&&PageID=3453&mode=2; [accessed 14.10.12].

[24] *www1.eere.energy.gov/geothermal/powerplants.html; [accessed 17.10.12].

[25] *http://en.wikipedia.org/wiki/Biomass; [accessed 17.10.12].

[26] *http://en.wikipedia.org/wiki/Ethanol; [accessed 17.10.12].

[27] *http://en.wikipedia.org/wiki/Ethanol_fuel; [accessed 17.10.12].

[28] *www.energyquest.ca.gov/story/chapter10.html; [accessed 17.10.12].

[29] *www.ucsusa.org/clean_energy/technology_and_impacts/energy_technologies/how-biomass-energy-works.html; [accessed 17.10.12].

[30] *What Are the Advantages of Biodiesel? www.k12.nf.ca/laval/ads.htm; [accessed 13.06.04].

[31] *National Biodiesel Board. www.biodiesel.org; [accessed 14.06.04].

[32] *www.nps.gov/renew/NPSBiodiesel.xls; [accessed 14.06.04].

[33] *Fact Sheet, 2001. www.eere.energy.gov/cleancities/blends/pdfs/biodiesel_fs.pdf; [accessed 14.06.04].

[34] *http://en.wikipedia.org/wiki/Methanol; [accessed 12.10.12].

[35] *www.afdc.energy.gov/afdc/ethanol/feedstocks_cellulosic.html; [accessed 12.10.12].

[36] *http://en.wikipedia.org/wiki/Solar_thermal_energy; [accessed 12.10.12].

[37] *http://en.wikipedia.org/wiki/Solar_power_plants_in_the_Mojave_Desert; [accessed 12.10.12].

[38] *http://peakenergy.blogspot.com/2008/04/concentrating-on-important-things-solar.html; [accessed 12.10.12].

[39] *www.greenoptimistic.com/2008/02/25/how-solar-parabolic-dish-concentrators-work/; [accessed 12.10.12].

[40] *www.rise.org.au/info/Tech/lowtemp/ponds.html.

[41] *www.solarserver.de/solarmagazin/artikel_april2001-e.htm; [accessed 12.10.12].

[42] *www.jc-solarhomes.com/fair/solar_greenhouse.htm; [accessed 12.10.12].

[43] *Department of Energy, Solar Trough Power Plant. Sun Lab: DOE. www.eren.doe.gov/sunlab; February 1999 [June 23, 2002].

[44] *Department of Energy, Solar Two Demonstrates Clean Power for the Future. Sun Lab: DOE. *www.eren.doe.gov/sunlab; March 2000 [accessed 23.06.02].

[45] *Department of Energy, Big Solutions for Big Proble, Concentrating Solar Energy. Sun Lab: DOE. *www.eren.doe.gov/csp; September 2001. [accessed 23.06.02].

[46] Solar Developments, Solar: Solar Thermal: Making Electricity from the Sun's Heat. www.solardev.com/SEIA-makingelec.php; [accessed 23.06.02].

[47] Department of Energy, Overview of Solar Thermal Technologies. Sun Lab: DOE. www.eren.doe.gov/csp; February 1998 [accessed 23.06.02].

[48] www.cader.org/microturbines.html; [accessed 15.10.12].

[49] www.globalmicroturbine.com/Site/Microturbine/Microturbine.html; [accessed 15.10.12].

[50] Space Studies Institute. http://ssi.org/solar-power-satellites/solar-power-satellite-art/; [accessed 19.10.12].

[51] www.docstoc.com/docs/3625467/Space-Based-Solar-Power; [accessed 19.10.12].

[52] Elsevier, Acta Astroautica. www.sciencedirect.com/science/article/pii/S0094576504001973; [accessed 19.10.12].

[53] Glaser PE. The future of power from the sun. US House of Representatives; May 7, 1973.

[54] *Satellite Power System Concept Development and Evaluation Program Reference System Report, US Department of Energy. http://chview.nova.org/station/sps.htm; [accessed 30.11.12].

[55] www.see.murdoch.edu.au/resources/info/Tech/wave/index.html; [accessed 19.10.12].

[56] www.thegga.org/50years-1970.htm; [accessed 19.10.12].

[57] www.ars.usda.gov/is/kids/transportation/story4/biodiesel.tw.sfk.htm; [accessed 30.11.12].

[58] http://solareis.anl.gov/guide/solar/csp/index.cfm; [accessed 19.10.12].

[59] http://bittooth.blogspot.com/2010_12_01_archive.html; [accessed 19.10.12].

[60] http://farmpolicy.com/wp-content/uploads/2012/01/NASS-CORN-12JAN.jpg; [accessed 19.10.12].

[61] http://earlywarn.blogspot.com/2011/02/us-ethanol-production.html; [accessed 19.10.12].

[62] http://respectrefugeesblog.org/category/charity/; [accessed 20.10.12].

[63] www.greentechnolog.com/2007/08/bagasse_as_biofuel_feedstock.html;   [accessed 20.10.12].

[64] www.nrel.gov/data/pix/searchpix.php?getrec=06442&display_type=verbose&search_reverse=1; [accessed 21.10.12].

[65] www.mindsetsonline.co.uk/tidalpower.php; [accessed 21.10.12].

[66] www.nrdc.org/energy/renewables/biomass.asp; [accessed 21.10.12].

[67] www.nrs.fs.fed.us/sustaining_forests/conserve_enhance/bioenergy/poplar_siting_land_use/; [accessed 21.10.12].

[68] www.seco.cpa.state.tx.us/energy-sources/biomass/forests.php; [accessed 21.10.12].

[69] http://maforests.org/; [accessed 21.10.12].

[70] www.mindsetsonline.co.uk/tidalpower.php; [accessed 21.10.12].

[71] www.ggsnews.com/13382/world-news-updates/major-tidal-energy-scheme-set-for-waters-off-islay/; [accessed 21.10.12].

[72] http://faracoseconsultancy.blogspot.com/2011/03/water-turbines-power.html; [accessed 21.10.12].

[73] www.window.state.tx.us/privacy.html; [accessed 22.10.12].

[74] http://inhabitat.com/oregon-wave-power/; [accessed 22.10.12].

[75] http://people.bath.ac.uk/mh391/WavePower/saltersduck.html; [accessed 22.10.12].

[76] www.jc-solarhomes.com/passive_solar.htm; [accessed 22.10.12].

[77] www.daviddarling.info/encyclopedia/S/AE_solar_hot_water_system.html.

[78] http://cpcbenvis.nic.in/newsletter/alternatefuel/ch140403.htm.

[79] www.caddet-re.org/html/body_299art4.htm; [accessed 31.10.12].

[80] www.montereybayaquarium.org/animals/AnimalDetails.aspx?enc=n3f4wmcSJaNrKzCzfXRa+Q==; [accessed 08.11.12].

[81] www.energy.kth.se/compedu/AsiaLink/S9_Sustainable_Technology/B1_Energy_Engineering/C5_Renewable_Energy/S9B1C5_files/Earth_interior_structure_and_its_dynamics.htm; [accessed 08.11.12].

Index

Note: Page numbers followed by *b* indicate boxes, *f* indicate figures, and *t* indicate tables.

Printed and bound by CPI Group (UK) Ltd, Croydon, CR0 4YY

14/10/2024

01773763-0001